Optimization in Electrical Engineering

Mohammad Fathi • Hassan Bevrani

Optimization in Electrical Engineering

Springer

Mohammad Fathi
University of Kurdistan
Kurdistan, Iran

Hassan Bevrani
University of Kurdistan
Kurdistan, Iran

ISBN 978-3-030-05308-6 ISBN 978-3-030-05309-3 (eBook)
https://doi.org/10.1007/978-3-030-05309-3

Library of Congress Control Number: 2018965419

This Springer imprint is published by the registered company Springer Nature Switzerland AG.
The registered company address is: Gewerbestrasse 11, 6330 Cham, Switzerland

Preface

There are a number of books available in the market describing electrical engineering optimization concepts. Most are primarily concerned with the mathematical theorems, which might not be interesting for students studying engineering.

The main purpose of this textbook is to provide engineering students with a comprehensive coverage of required fundamentals. Numerous applications and examples are provided to demonstrate how the discussed fundamentals can be utilized in the domain of engineering, with a particular focus on electrical engineering.

It is expected that students, academic researchers, engineers, and institutes in engineering will find this book useful, particularly those in the area of electrical engineering. This book can be used as the main text in courses such as Engineering Optimization, Convex Optimization, Advanced Engineering Mathematics, and Robust Optimization. The book can also be utilized as a supplementary text for university students at both undergraduate and postgraduate levels for courses such as Smart Grids, Communication Networks, Wireless Communication, Automation, and Systems Theory. The main prerequisite of optimization is linear algebra; to satisfy this requirement, the book includes an extensive chapter on this topic.

The authors have over a decade of experience in both academia and industry, specifically with communication networks/systems, automation, robust/intelligent control, and smart/microgrids. The application of such experience to this text creates a rich and detailed resource for students, researchers, and engineers studying/working on electrical engineering.

Most of contributions, outcomes, and insights presented in this book were achieved through long-term teaching and research conducted by the authors and their research groups on optimization problems over the years. It is a pleasure to acknowledge the received supports and awards from the Smart/Micro Grids Research Center (SMGRC) and the University of Kurdistan (UOK) as well as other sources. In particular, the comments from our colleague Dr. S. Fathi Manesh at the UOK are appreciated.

The authors would like to thank their postgraduate students, S. Shokoohi, S. Hasani, Y. Khayat, A. Abdollahi, S. Rehimi, and M. Abdolmaleki, as well as the students in the courses of Advanced Engineering Mathematics, Smart Grids, and Robust Optimal Control in the Department of Electrical Engineering at the UOK for their active and continuous support.

Last but not least, the authors offer their deepest personal gratitude to their families for all their patience and help during the preparation of this book.

October 2018

Mohammad Fathi
Hassan Bevrani

Contents

1 Introduction

Abstract

Over the years, making the best decision among a number of possible choices has been a challenging issue. The issue arises from the complexity of the decision criteria and the extent of possible choices. Optimization is the knowledge of decision-making. Scientists and engineers have always tried to develop techniques and tools to overcome the complexity of optimization.

Keywords

Optimization in electrical engineering · Signal processing · Circuit design · Resource allocation · Power dispatching · Load-frequency control · Microgrid planning

1.1 Optimization Taxonomy

In the area of engineering, the first step towards optimization is to model the decision-making process mathematically into an optimization problem. Depending on the structure of the decision criteria and possible choices, optimization problems are generally categorized into linear and nonlinear programming problems [1]. Deployed functions in a linear program are limited to appear in the linear form. Reliable and efficient algorithms along with implemented solvers have been developed for these problems [2]. Consequently, linear programming is a mature area and mostly considered as a technology. In spite of the strict limitation in the form of deployed functions, linear programming has wide applications in engineering.

On the other hand, the area of nonlinear programming remains a challenging domain due the diversity and complexity in the form of functions deployed in the problem [3]. Nonlinear programming in turn can be divided into different categories. Among which, integer programming [4, 5], convex programming [6, 7], and geometric programming [8] are more common due to the solution procedures proposed in the literature.

Modeling a problem as a nonlinear programming is relatively easy in general. This is due to the fact there is no limitation in the form of functions and parameters deployed in the problem. Indeed, they can appear in any nonlinear function and integer/continuous parameters. However, solving a nonlinear programming problem analytically is not trivial in general. Alternatively, artificial intelligence-based tools such as artificial neural network [9], particle swarm optimization [10], and genetic algorithm [11]

M. Fathi, H. Bevrani, *Optimization in Electrical Engineering*,
https://doi.org/10.1007/978-3-030-05309-3_1

have been investigated over the years. Inspired by the human and nature, these tools are mostly based on search methods. They employ intelligent techniques to explore the whole possible solutions of the given problem and therefore to overcome some limitations of traditional optimization approaches.

To mitigate the broadness of nonlinear programming and to explore analytic solution procedures, the focus is usually done on a number of special cases of nonlinear problems such as convex and geometric programmings, in which deployed functions are limited to appear in certain forms [12]. The challenging issue is in the modeling of a problem in these forms and moreover to recognize whether a given problem follows the required forms or not. In return of accepting these limitations, solving these programmings is possible through reliable and efficient algorithms [13, 14]. Consequently, the area of convex and geometric programming is mostly considered as a technology. Although there is no analytic solution in general, reliable and efficient algorithms implemented by software packages do exist for this area [15–17].

1.2 Optimization in Engineering

Engineering is the knowledge of analysis and design. The most important requirement of designing is the simplicity in the design and the optimality in the performance. In the past decades, with a small number of components in the process of decision-making, achieving these requirements together was almost readily available. However, nowadays, with a large number of components in engineering systems, achieving this goal is not trivial. An alternative tool towards this goal is optimization [18]. The area of engineering, particularly electrical engineering, has always been affected by mathematical tools and technologies. One of these tools that have emerged in recent years is mathematical optimization.

1.3 Optimization in Electrical Engineering

Over the years, the area of electrical engineering was about analysis and synthesis of a system with a limited or a small number of elements. Nowadays, things have been changed. Not only systems are getting larger and more complex but also they are connecting and collaborating together. They work towards a common objective while facing some constraints.

To address the requirements of emerging electrical engineering systems, this area is recently influenced by mathematical optimization [19, 20]. This is of high importance when there is a need that engineering systems must operate economically. The knowledge of optimization helps engineers to design more efficient systems. They would be able to carry out systematic modeling, designing, and analysis of complex networks and systems considering required constraints and parameters. For example, communication and electrical power infrastructures are getting more intelligent, a higher number of electronic devices are placed on integrated circuits, and more data is generated and processed by the information technology. Design and analysis in all of these areas need appropriate tools of optimization, which is the topic of this book.

As a motivation and to highlight the topic, a number of application examples of optimization in electrical engineering are stated in the following subsections.

1.3.1 Signal Processing

Many problems in signal processing can either be cast as or be converted into optimization problems. With significant advancements of optimization methods and computing resources, these problems can be solved efficiently with recently developed optimization models and software tools. More importantly, the advances in computing resources enable these optimization tools to operate in the real-time manner. This motivates the extension of innovative applications of optimization in the area of signal processing.

One known application of optimization in the domain of signal processing was the design of linear phase finite impulse response filters in which filter coefficients are optimized [21, 22]. In array signal processing, multiple sensor outputs are linearly combined together to form a composite array. The output weights are then optimized to form a desired pattern [23]. The parameters of classification algorithms in statistical and machine learning are determined with optimization methods [24]. Image and speech processing applications take advantage of optimization techniques for noise removal, recognition, and automation [25–27]. Parameter estimation and data fitting are also optimized using fitting moving average (MA) or autoregressive moving average (ARMA) models [28, 29]. The performance of many functional blocks in communication systems and networks including equalization, pulse shape design, demodulation, detection, antenna selection, compressed spectrum sensing, etc. is improved using optimization methods [30–32].

1.3.2 Circuit Design

With the development of VLSI technology, integration of mixed analog–digital circuits as a complete system on a chip is in progress. Although the analog part forms a small part of the entire circuit, its design is much more difficult due to the complex nature of analog circuit design and the trade-off between several number of parameters.

This design is often done in two stages [33]. In the first stage, the circuit topology and structure is constructed. Then, in the second stage, the circuit parameters such as the size of elements are determined in order to optimize performance measures such as area, gain, and consumption power. This stage that is traditionally done by hand using analytic equations is an iterative and time-consuming task without guarantying the optimal performance. Consequently, analog circuit design automation techniques become essential to obtain the optimal performance with the minimum time effort [34].

To overcome the aforementioned complexities and drawbacks, optimization methods have been employed by the researchers [35]. Due to a trade-off between several performance measures, design problems in this area mostly appear in the form of multi-objective optimization. Optimization methods and tools not only reduce the time of circuit design but also result in more efficient, robust, and cost-effective circuits.

Analog circuit design based on geometric programming has been presented in [36]. Applications of evolutionary optimization methods in the design of analog integrated circuits have been surveyed in [37]. A particle swarm optimization method is used in [38] towards multi-objective circuit design and extracting Pareto feasible points. Parameter optimization and device sizing using evolutionary algorithms have also been presented in [39] and [40]. Furthermore, the implementation of radio frequency energy harvesting circuits is also done using optimization methods in [41].

1.3.3 Resource Allocation in Communication Networks

Due to the diversity in the number of supported users and services, communication networks have demonstrated their importance and progressing role for the community over the years. The performance of these networks is highly impacted by the methodologies employed to design and to

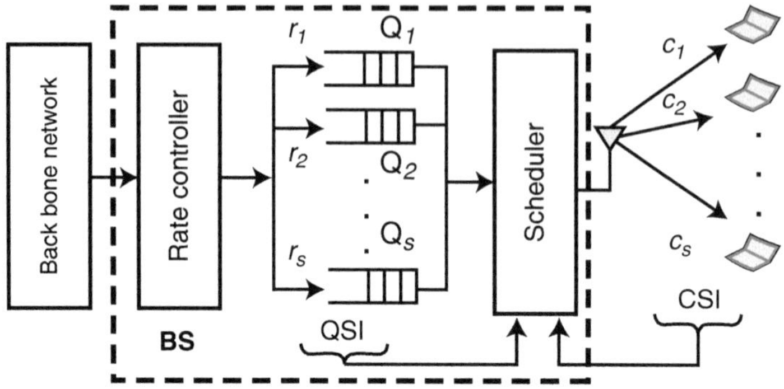

Fig. 1.1 Controller blocks in a communication network

allocate communication resources among the network entities [42]. The resources appear in a number of forms including time, geographical space, transmitted power by antennas, frequency spectrum provided by the network utility, buffer spaces in the network equipment, etc. The purpose of resource allocation in communication networks is to intelligently assign the limited available resources among the network entities in an efficient way to satisfy the user's service requirements [43].

With the fast development of communication networks and diversity in the number of transmission technologies, resource allocation becomes a challenging issue. In particular, complexities in the network topology, different radio technologies in wireless communication, environmental concerns, diversity in user requirements, and growing demand for high-speed data transmission cause the resource allocation to be more critical. To satisfy this requirement, a number of tools including optimization theory, control theory, game theory, and intelligent methods have been employed recently to model and to design resource allocation methodologies in communication networks [44–46]. Due to the dynamic nature of communication parameters such as the communication channels and the transmitted traffic, these methodologies need to be of low computational complexity to operate in a real-time manner.

A typical wireless communication network consisting of a base station (BS) and a number of receivers is shown in Fig. 1.1 [47]. Associated with each receiver, there is a queue to buffer arrival data packets by a block known as the rate controller. The packets are then scheduled to be transmitted by a radio known as the scheduler. The scheduler decision is mostly based on the network quality of service requirements and the provided information such as the queue state information (QSI) and channel stated information (CSI). The performance of the network in terms of the joint performance of the rate controller and the scheduler is usually formulated into an optimization problem, known as the network utility maximization. The problem is then solved by decomposing it into individual subproblems in the rate controller and scheduler blocks.

The resource allocation in communication networks has been addressed by many research publications recently using optimization methodologies. Rate control of data transmitters has been used as a way to combat congestion in communication networks [48]. Allocation of resources among different protocol layers in wireless networks has been done using an approach known as cross-layer design [49, 50]. Joint rate control, routing, and scheduling in wireless networks in different radio technologies has been decoupled into distinct problems in different protocol layers [46, 51]. In addition to proposing general frameworks of resource allocation in wireless communication, each radio technology and topology in wireless networks can also be investigated in more detail from this viewpoint, e.g., cellular networks [52], device-to-device communication [53], and wireless sensor networks [54].

1.3.4 Power Dispatching

Power generation units in electric power systems face a number of constraints such as fuel, generation levels, ramping rate, output uncertainties, and the system load demand. The process of power dispatching is to determine and to allocate generation levels to generation units in order to satisfy the load and power demand in the network while considering the units and the network constraints. Mostly, the objective in power dispatching is to minimize the cost of power production, which is known as economic power dispatching problem [55]. Due to the constraints of the network and generation units, transmission losses, and generation and load uncertainties, power dispatching problem has been a challenging issue for decades [56].

To overcome the complexity of power dispatching, optimization and control tools have been employed widely by the researchers [57]. Intelligence-based optimization tools have been employed widely for dispatching problems [58,59]. The uncertainty from both sides of generation and demand in next-generation power grids needs new techniques on stochastic modeling and optimization [60, 61]. A comprehensive discussion of deterministic and stochastic optimization techniques towards optimal power flow, with an emphasis on the advantages, disadvantages, and computational characteristics of each technique, is presented in [62] and [63].

Power dispatching problem is even more critical in next-generation power grids consisting of new blocks, known as microgrid (MG) [64]. An MG is a small electrical distribution system that interconnects multiple customers, distributed generators (DGs), renewable energy sources, and storage energy systems (ESSs). The MGs are known as the main building blocks of the future smart grids and are typically characterized by multipurpose electrical power services to communities that are connected via low-voltage networks. Due to the capability of MGs to operate in an interconnected mode, power dispatching problem in MGs is mostly addressed by establishing a set of power flows among MGs in the literature [61], as shown in Fig. 1.2. Moreover, MGs can contribute to regulating power management by injecting their surplus active/reactive powers. The network operator has the responsibility of managing the regulating power [65]. To ease the implementation of power dispatching and regulating in interconnected MGs, wireless communication infrastructures and protocols have been proposed in [66] and [67].

1.3.5 Optimal Load-Frequency Control

Load-frequency control (LFC) loop as the main part of automatic generation control (AGC) in power systems is responsible to minimize the system frequency deviation and balance the grid load and

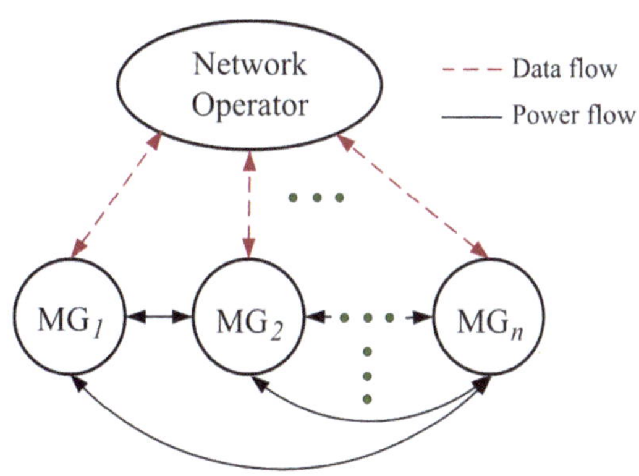

Fig. 1.2 Power flow in interconnected MGs

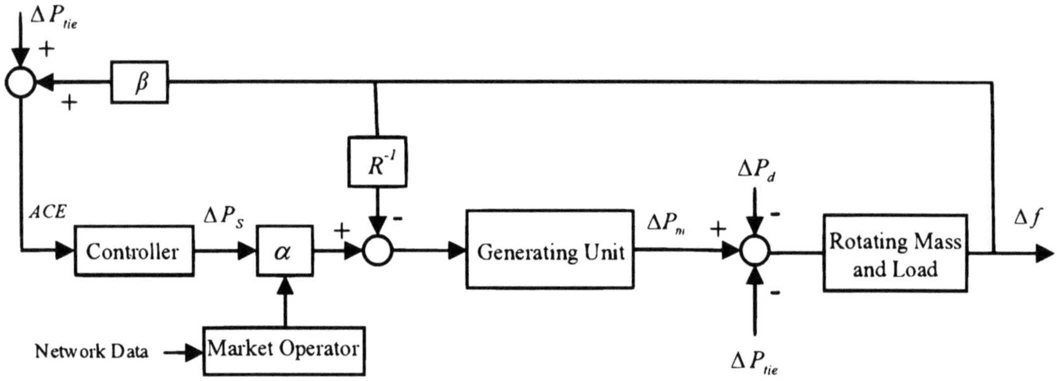

Fig. 1.3 Conceptual LFC response model

generated power. In addition to minimizing the frequency change in a control area, the LFC must maintain the power interchanges at the scheduled values with other connected control areas. These objectives are satisfied by measuring the area control error (ACE) signal representing the real power imbalance between power generation and load via the following linear combination of net power interchange and frequency deviation (for typical area i) [68]:

$$ACE_i = \Delta P_{tie,i} + \beta_i \Delta f_i \tag{1.1}$$

where $\Delta P_{tie,i}$ is tie-line power change, Δf_i is the area frequency deviation, and β_i is the area bias factor.

A conceptual LFC response model is shown in Fig. 1.3. The ΔP_m is the generator mechanical power change, R is the speed droop characteristic showing the speed regulation due to governor action, and ΔP_d is the load/generation disturbance. The ΔP_s is the (LFC or secondary) control action signal. The α is participation factor of generating unit in the LFC. Market operator is responsible to balance the system generation-load in a reliable, secure, and economic way. Market operator can change the bulk generators setpoint, participation factor, and power dispatching through the LFC and higher control level. These parameters are fully explained in [68, 69].

The LFC synthesis problem in a multi-area power system can be easily solved using the following optimization problem over all controllers $k_j \in K_S$ (K_S is set all feasible stabilizing controllers), considering a constraint on the control action signal, with the upper limit u_0:

$$\min_{k_j \in K_S} \; |ACE_i|, \;\; \forall i \tag{1.2a}$$

$$\text{subject to } \; |\Delta P_s| < u_0 \tag{1.2b}$$

1.3.6 Optimal Operational Planning in Microgrid

In order to allow seamless deployment of MGs, the relevant operation and control challenges should be solved. Consider an MG with several renewable energy sources such as photovoltaic (PV) and wind turbine (WT) units, ESSs, and conventional DGs (CDGs) such as diesel engine and micro-gas turbines (MGT). In order to follow an optimal operational planning in this MG, the environmental and economic issues must be taken into account by the MG control center (MGCC) [70]. For this purpose, two stages are usually required. First stage is a day-ahead operational planning algorithm for the minimization of CO_2 emissions and fuel consumption by the CDGs. This algorithm may solve the unit commitment problem (UCP) (e.g., using a dynamic programming) with predictions of the available energy from PV and WT generators, the power demand from the loads, and the state of charge (SOC) of ESSs. Based on these data, this stage gives power references for CDGs while maximizing the use of PV and WT generators and reducing equivalent CO_2 emissions and the economic operating cost.

The second stage may be implemented during the day and consists in reducing variations due to the power uncertainty (from the PV/WT production and load demand). This stage enables to retrieve day-ahead calculated optimal economic costs and/or CO_2 emissions. Ensure uninterruptable power supply to the loads, maximize renewable generation in the energy mix, and minimize the economic costs and the CO_2 equivalent emissions of the CDGs. The later task can be done by setting the CDGs power references such that they produce the minimum pollution and have minimum startups and shutdowns are important objectives of the MGCC. A general scheme to formulate and solve the optimal operational planning problem for cost minimization and CO_2 equivalent emissions reduction is depicted in Fig. 1.4.

The dynamic programing-based optimization procedure can be performed with three different objective functions: minimization of the CO_2 equivalent emissions from the CDGs, the consumed fuel and a trade-off between these two functions. In [70], the objective function of UCP with dynamic programming is applied to minimize the CO_2 emissions and cost according to the given operational planning scheme in Fig. 1.4. To solve the objective function, different power and reserve dispatching strategies are performed under several nonlinear constraints.

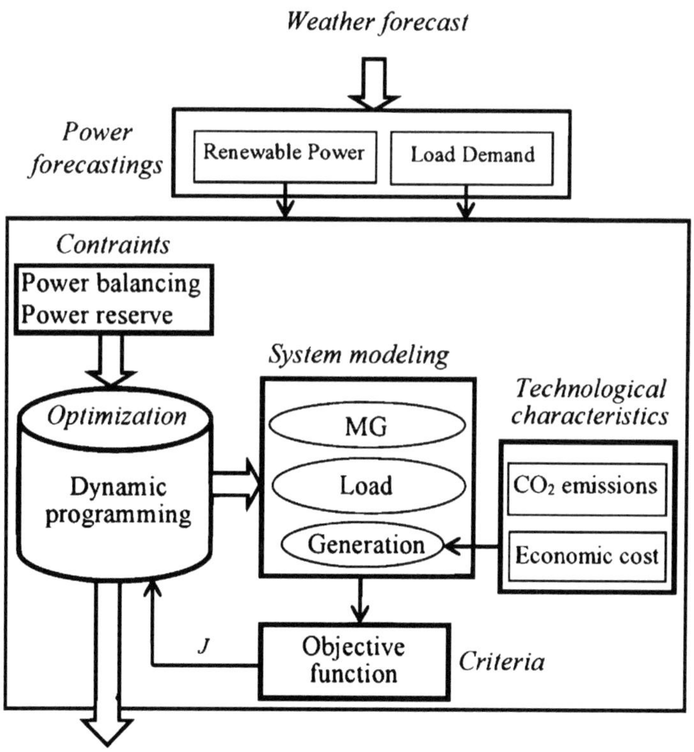

Fig. 1.4 Scheme of the day-ahead optimal operational planning

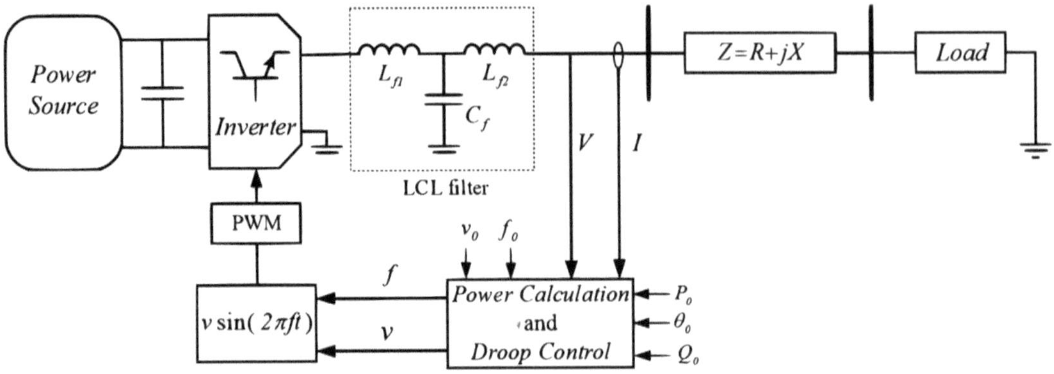

Fig. 1.5 An inverter-based DG with droop control

1.3.7 Microgrid Voltage–Frequency Control

An MG may include several DGs. Figure 1.5 shows an example of an inverter-based DG for supplying a local load through a line in an MG. The DG is connected to the load bus via a line impedance $Z\angle\theta$. $Z = R + jX$ is the interconnection line impedance and θ is the impedance angle. The LCL output filter has been added to prevent the resonance impact in the output network. Also, the LCL damps the distortion of output sinusoidal waveform and reduces high-frequency harmonics caused by switching operations of the inverter [70].

In Fig. 1.5, f_0 and v_0 are, respectively, rated as nominal frequency and voltage of the MG. P_0 and Q_0 are the nominal values of active and reactive powers, respectively. If a change happens in the frequency or voltage of inverter, the impact can be observed on the output active and reactive powers of the inverter. The amount of suitable frequency and voltage deviation can be achieved through the droop characteristics. The calculated powers are used by the *Droop Control* block to provide corresponded voltage and frequency deviations. The droop control stabilizes the DG's frequency and voltage following load changes.

It is shown in [71] that by properly estimating the line parameters, the droop control would exhibit high performance and acceptable response at different load change scenarios. For this purpose, an objective function must be determined to minimize the frequency and voltage fluctuations. For example, simultaneous control of frequency and voltage can be presented in the following optimization problem:

$$\min \sum_{i}^{n} \alpha_i |\Delta f_i(x)| + \beta_i |\Delta v_i(x)| \tag{1.3a}$$

$$\text{subject to } |\Delta f_i| < E_f,\ |\Delta v_i| \le E_v,\ R_i^{\min} \le R_i \le R_i^{\max}, X_i^{\min} \le X_i \le X_i^{\max},\ \forall i \tag{1.3b}$$

where x is the line resistance (R) and reactance (X) of DGs, $f(x)$ is the objective function, Δf_i is the frequency deviation of the ith DG output, Δv_i the voltage deviation of the i-th DG output, α_i is the stress coefficient of frequency in node i, β_i is the stress coefficient of voltage in node i, n is the total number of DGs in the MG, E_f is the maximum allowable frequency deviation, and E_v is the maximum allowable voltage deviation. For optimal estimation of line parameters, the particle swarm optimization (PSO) evolutionary algorithm has been used in [68], and the obtained values are tested on an MG case study. In [71], to solve the same optimization problem in an MG with numerous DGs, an adaptive neuro-fuzzy inference system is used instead of the droop control.

1.3.8 Wide-Area Power System Oscillation Damping

Wide-area power system oscillation damping using wide-area measurement system (WAMS) with phasor measurement units (PMUs) is an important issue in a modern electric power system design and operation. An accurate monitoring can be achieved by the technique of phasor measurements and global positioning system (GPS) time signal. Proper grasp of the present state with flexible wide area control and smart operation address significant elements to maintain wide-area stability in the complicated grid with the growing penetration of RESs/DGs and MGs [72].

Using PMU-based WAMS, the parameters of the power system stabilizers (PSSs) and/or automatic voltage regulators (AVRs) can be adjusted using an on-line monitoring-based optimal tuning mechanism. This control architecture is conceptually shown in Fig. 1.6. This figure shows the schematic diagram of the concept of optimal damping controller based on the wide-area phasor measurements. The small signal stability assessment with phasor measurements is well discussed in [72]. The stability of the inter-area low-frequency oscillation mode can be investigated by adopting the method to estimate the oscillation dynamics with a simple oscillation model. The estimated low-order model based on the measurements represents power system oscillations, and can be used to design/tune PSSs.

To stabilize the dominants oscillation modes, the specified PSSs (and/or AVRs) in a power system can be tuned based on the estimated low-order oscillation model. The effectiveness of tuning approach can be assessed more properly by evaluating the eigenvalues of estimated oscillation model which is obtained from the measurement data. One may tune the damping controllers gradually, and in each step, the effect of tuning is assessed by eigenvalues of the low-order oscillation model that is estimated again using the phasor fluctuations after tuning control.

In [72], a coupled vibration model with two dominant electro-mechanical modes (a dominant and a quasi-dominant modes) associated with the angle stability is discussed. The result shows that the approximated model estimates two conjugate eigenvalues, properly. In addition, the trend of the

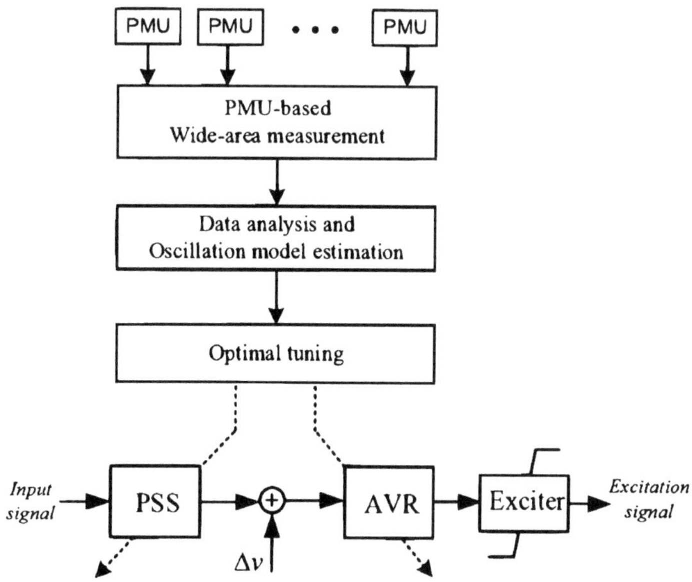

Fig. 1.6 A structure for wide-area optimal tuning of PSS-AVR systems

variation of both eigenvalues with parameter change coincides well with each other. Therefore, the extended coupled vibration model can correctly evaluate the stability change with tuning of PSS parameters.

The PSSs parameters are determined to improve damping ratios of two low-frequency modes of the model, that is, to maximize the following objective function J under the constraint that all eigenvalues lie on the left-hand side of the complex plane:

$$\max \quad J = \min_{i} \frac{\Re(\lambda_i)}{\sqrt{\Re(\lambda_i)^2 + \Im(\lambda_i)^2}} \tag{1.4a}$$

$$\text{subject to } \Re(\lambda_i) < 0, \quad \forall i \tag{1.4b}$$

where λ_i is the i-th eigenvalue of estimated low-order oscillation model. The $Re(\lambda_i)$ and $\Im(\lambda_i)$ denote the real and imaginary parts of the i-th eigenvalue, respectively.

1.4 Purpose and Structure

This book aims to provide students, researchers, and engineers in the area of electrical engineering with a textbook in mathematical optimization. Using a readable wording without getting involved into the details of mathematical theorems, the book tries to highlight fundamental concepts of optimization used in electrical engineering. From simple but important concepts in optimization such as unconstrained optimization to highly advanced topics such as linear matrix inequalities and artificial intelligence-based optimization methodologies have been collected together in the book.

The reader is motivated to be engaged with the content via numerous application examples of optimization in the area of electrical engineering throughout the book. This approach not only provides relevant readers with a better understanding of mathematical basics but also reveals a systematic approach of algorithm design in the real world of electrical engineering. The content of the book is so arranged and ordered that anyone with a basic knowledge in mathematics can follow the chapters.

Overall, the book contains specific and useful optimization materials in engineering. University students and academic researchers in electrical engineering are primary audiences of the proposed book. Indeed, the book can be used as a textbook for this society. Moreover, the book can be used as a supplement book by university students, design engineers, and academic researchers in other areas of engineering.

1.5 Organization

The book proceeds with an extended review of linear algebra in Chap. 2 that is a prerequisite to the mathematical optimization. This chapter provides an intensive review of linear algebra with the aim of covering mathematical fundamentals in the following chapters. It discusses vector and matrix spaces, and their application in analyzing the solution of a set of linear equations. The chapter then discusses well-known quadratic forms and introduces positive and negative definite properties accordingly. Matrix diagonalization and decomposition is presented using eigenvalue and singular value definitions. Then, some calculus concepts such as level set, gradient, and Hessian are finally presented.

The concept of optimization including objective function, optimization variable, feasible set, and optimal value is introduced in Chap. 3. This is done using the set constrained optimization as the simplest form of optimization. The chapter then discusses some conceptual optimality conditions for the set constrained optimization. A number of basic and general iterative optimization algorithms are then introduced with extensive examples.

Chapter 4 discusses convex programming as a category with a wide range of applications in all fields of engineering. This kind of programming is characterized by some conditions and properties in the construction of functions used in the objective and constraint functions. One major advantage of convex programming is that any local optimal point is also global, which brings forward a great step in the solution of convex optimization problems. The chapter first defines convex sets and convex functions, using which the definition of a convex programming problem is determined. To clarify these definitions, appropriate application examples in electrical engineering are given accordingly.

Solving an optimization problem in general is difficult. Chapter 5 discusses the concept of duality as a framework to transfer the problem into another domain, known as dual domain. Based on the achievements from this domain, Karush–Kuhn–Tucker (KKT) conditions are introduced, which are helpful in finding optimal solutions in general optimization problems. Some application examples in electrical engineering are given, accordingly.

Chapter 6 presents the application of linear matrix inequalities (LMIs) in real-world engineering optimization problems in both convex and non-convex categories. The LMI is used as a powerful mathematical language for reformulation of a wide range of optimization problems. As examples, some robust optimization problems are formulated and solved via the LMI-based optimal H_∞ and mixed H_2/H_∞ control theorems. In order to solve non-convex optimization problems, iterative LMI algorithms are introduced. The chapter is supplemented by several illustrative examples and MATLAB codes.

Chapter 7 shows that how using the artificial intelligence (AI) and nature-inspired searching methods relaxes some constraints as well as limitations of well-known optimization approaches, and increases the number of solvable optimization problems. The application of artificial neural networks (ANNs), particle swarm optimization (PSO), and genetic algorithm (GA) for solving the practical optimization problems is explained. The key ideas and several application examples are presented. The flexibility and simplicity of AI and evolutionary-based algorithms in solving of complex multi-objective optimization problems are emphasized.

References

1. D.G. Luenberger, Y. Ye, *Linear and Nonlinear Programming* (Springer, Berlin, 2016)
2. M.S. Bazaraa, J.J. Jarvis, H.D. Sherali, *Linear Programming and Network Flows* (Wiley, Hoboken, 2009)
3. D.P. Bertsekas, *Nonlinear Programming* (Athena Scientific, Belmont, 1999)
4. L.A. Wolsey, *Integer Programming* (Wiley, Hoboken 1998)
5. H.M. Salkin, K. Mathur, *Foundation of Integer Programming* (North-Holland, New York, 1989)
6. D.P. Bertsekas, E. Nedic, A. Ozdaglar, *Convex Analysis and Optimization* (Athena Scientific, Nashua, 2003)
7. S. Boyd, L. Vandenberghe, *Convex Optimization* (Cambridge University Press, Cambridge, 2004)
8. R. Duffin, E. Peterson, C.Zener, *Geometric Programming–Theory and Application* (Wiley, New York, 1967)
9. A.I. Galushkin, *Neural Networks Theory* (Springer, New York, 2007)
10. J. Kennedy, R. Eberhart, Particle swarm optimization. *Proceedings of the IEEE International Conference on Neural Networks*, vol. 4 (IEEE, Perth, 1995), pp. 1942–1948
11. F. Rothlauf, *Representations for Genetic and Evolutionary Algorithms* (Springer, New York, 2006)
12. S. Boyda, S.J. Kim, L. Vandenberghe, A. Hassibi, A tutorial on geometric programming. Optim. Eng. **8**(1), 67–127 (2007)
13. Y. Nesterov, A. Nemirovsky, *Interior Point Polynomial Algorithms in Convex Programming* (SIAM, Philadelphia, 1994)

14. K.O. Kortanek, X. Xu, Y. Ye, An infeasible interior-point algorithm for solving primal and dual geometric programs. Math. Program. **76**(1), 155–181 (1996)
15. M. Grant, S. Boyd, CVX: MATLAB software for disciplined convex programming, version 2.1. (2014) http://cvxr.com/cvx
16. MOSEK, Mosek optimization toolbox. (2002). www.mosek.com
17. GNU linear programming kit, version 4.45. http://www.gnu.org/software/glpk
18. S.S. Rao, *Engineering Optimization: Theory and Practice* (Wiley, Hoboken, 2009)
19. M. Baker, *Nonlinear Optimization in Electrical Engineering with Applications in MATLAB* (IET, Hertfordshire, 2013)
20. J.A. Taylor, *Convex Optimization of Power Systems* (Cambridge University Press, Cambridge, 2015)
21. L. Rabiner, Linear program design of finite impulse response (FIR) digital filters. IEEE Trans. Audio Electroacoust. **20**(4), 280–288 (1972)
22. S.-P. Wu, S. Boyd, L. Vandenberghe, FIR filter design via spectral factorization and convex optimization. *Applied and Computational Control, Signals, and Circuits* (Springer, Berlin, 1999), pp. 215–245
23. H. Lebret, S. Boyd, Antenna array pattern synthesis via convex optimization. IEEE Trans. Signal Process. **45**(3), 526–532 (1997)
24. H. Trevor, R. Tibshirani, J. Friedman, *The Elements of Statistical Learning: Data Mining, Inference, and Prediction* (Springer, New York, 2009)
25. L.I. Rudin, S. Osher, E. Fatemi, Nonlinear total variation based noise removal algorithms. Phys. D: Nonlinear Phenom. **60**(1–4), 259–268 (1992)
26. W. Zhao, R. Chellappa, P. J. Phillips, A. Rosenfeld, Face recognition: a literature survey. ACM Comput. Surv. **35**(4), 399–458 (2003)
27. Y. Gong, Speech recognition in noisy environments: a survey. Speech Commun. **16**(3), 261–291 (1995)
28. P. Stoica, T. McKelvey, J. Mari, MA estimation in polynomial time. IEEE Trans. Signal Process. 48(7), 1999–2012 (2000)
29. B. Dumitrescu, I. Tabus, P. Stoica, On the parameterization of positive real sequences and MA parameter estimation. IEEE Trans. Signal Process. **49**(11), 2630–2639 (2001)
30. T.N. Davidson, Z.-Q. Luo, K.M. Wong, Design of orthogonal pulse shapes for communications via semidefinite programming. IEEE Trans. Signal Process. **48**(5), 1433–1445 (2000)
31. A. Dua, K. Medepalli, A.J. Paulraj, Receive antenna selection in MIMO systems using convex optimization. IEEE Trans. Wirel. Commun. **5**(9), 2353–2357 (2006)
32. G. Sell, M. Slaney, Solving demodulation as an optimization problem. IEEE Trans. Audio Speech Lang. Process. **18**(8), 2051–2066 (2010)
33. S.L. Sabat, K.S. Kumar, S.K. Udgata, Differential evolution and swarm intelligence techniques for analog circuit synthesis. *Proceeding of the World Congress on Nature and Biologically Inspired Computing* (IEEE, Coimbatore, 2009), pp. 469–474
34. G. Alpaydin, S. Balkir, G. Dundar, An evolutionary approach to automatic synthesis of high-performance analog integrated circuits. IEEE Trans. Evol. Comput. **7**(3), 240–252 (2003)
35. B. Liu, G. Gielen, F.V. Fernández, Automated design of analog and high-frequency circuits. *Proceeding of the Fundamentals of Optimization Techniques in Analog IC Sizing. Studies in Computational Intelligence* (Springer, Berlin, 2014)
36. S. Boyda, S.J. Kim, D. Patil, M. Horowitz, Digital circuit optimization via geometric programming. Oper. Res. **53**(6), 899–932 (2005)
37. E. Tlelo-Cuautle, I. Guerra-Gómez, M.A. Duarte-Villaseñor, Applications of evolutionary algorithms in the design automation of analog integrated circuits. J. Appl. Sci. **10**(17), 1859–1872 (2010)
38. M. Fakhfakh, Y. Cooren, A. Sallem, M. Loulou, P. Siarry, Analog circuit design optimization through the particle swarm optimization technique. Analog Integr. Circuits Signal Process. **63**(1), 71–82 (2010)
39. P.P. Kumar, K. Duraiswamy, A.J. Anand, An optimized device sizing of analog circuits using genetic algorithm. Eur. J. Sci. Res. **69**(3), 441–448 (2012)
40. D. Nam, Y.D. Seo, L.J. Park, C.H. Park, B. Kim, Parameter optimization of an on-chip voltage reference circuit using evolutionary programming. IEEE Trans. Evol. Comput. **5**(4), 414–421 (2001)
41. P. Nintanavongsa, U. Muncuk, D.R. Lewis, K.R. Chowdhury, Design optimization and implementation for RF energy harvesting circuits. IEEE J. Emerging Sel. Top. Circuits Syst. **2**(1), 24–33 (2012)
42. D. Bertsekas, R.G. Gallager, *Data Networks* (Prentice Hall, New Jersey, 1991)
43. Y. Su, F. Fu, S. Guo, Resource allocation in communications and computing. Can. J. Electr. Comput. Eng. **2013**, 328395 (2013)
44. A. Ephremides, S. Verdu, Control and optimization methods in communication network problems. IEEE Trans. Autom. Control **9**(1), 930–942 (1989)

45. Z. Han, K.J.R. Liu, *Resource Allocation for Wireless Networks: Basics, Techniques, and Applications* (Cambridge University Press, Cambridge, 2008)
46. M. Fathi, H. Taheri, M. Mehrjoo, Cross-layer joint rate control and scheduling for OFDMA wireless mesh networks. IEEE Trans. Veh. Technol. **59**(8), 3933–3941 (2010)
47. M. Fathi, H. Taheri, M. Mehrjoo, Utility maximisation in channel-aware and queue-aware orthogonal frequency division multiple access scheduling based on arrival rate control. IET Commun. **6**(2), 235–241 (2012)
48. F. Kelly, Charging and rate control for elastic traffic. Eur. Trans. Telecommun. **8**(1), 33–37 (1997)
49. W. Stanczak, M. Wiczanowski, H. Boche, *Fundamentals of Resource Allocation in Wireless Networks: Theory and Algorithms* (Springer, Berlin, 2008)
50. S. Shakkottai, T.S. Rappaport, P.C. Karlsson, Cross-layer design for wireless networks. IEEE Commun. Mag. **41**(10), 74–80 (2003)
51. M. Chiang, S.H. Low, A.R. Calderbank, J.C. Doyle, Layering as optimization decomposition: a mathematical theory of network architectures. Proc. IEEE **95**(1), 255–312 (2007)
52. M. Fathi, E. Karipidis, Distributed allocation of subcarrier, power and bit-level in multicell orthogonal frequency-division multiple-access networks. IET Commun. **8**(6), 781–788 (2014)
53. S. Sharifi, M. Fathi, Underlay device to device communication with imperfect interference channel knowledge. Wirel. Pers. Commun. **101**(2), 619–634 (2018)
54. M. Fathi, V. Maihami, Operational state scheduling of relay nodes in two-tiered wireless sensor networks. IEEE Syst. J. **9**(3), 686–693 (2015)
55. B. Chowdhury, S. Rahman, A review of recent advances in economic dispatch. IEEE Trans. Power Syst. **5**(4), 1248–1259 (1990)
56. M. Huneault, F.D. Galiana, A survey of the optimal power flow literature. IEEE Trans. Power Syst. **6**(2), 762–770 (1991)
57. J. Zhu, *Optimization of Power System Operation* (Wiley-IEEE Press, Hoboken, 2015)
58. M.A. Abido, Environmental/economic power dispatch using multiobjective evolutionary algorithms. IEEE Trans. Power Syst. **18**(4), 1529–1537 (2003)
59. M.R. AlRashidi, M.E. El-Hawary, A survey of particle swarm optimization applications in electric power systems. IEEE Trans. Evol. Comput. **13**(4), 913–918 (2009)
60. M. Fathi, H. Bevrani, Adaptive energy consumption scheduling for connected microgrids under demand uncertainty. IEEE Trans. Power Deliv. **28**, 1576–1583 (2013)
61. M. Fathi, H. Bevrani, Statistical cooperative power dispatching in interconnected microgrids. IEEE Trans. Sustain. Energy **4**, 586–593 (2013)
62. S. Frank, I. Steponavice, S. Rebennack, Optimal power flow: a bibliographic survey I. Energy Syst. **3**(3), 221–258 (2012)
63. S. Frank, I. Steponavice, S. Rebennack, Optimal power flow: a bibliographic survey II. Energy Syst. **3**(3), 259–289 (2012)
64. O. Gandhi, C.D. Rodríguez-Gallegos, D. Srinivasan, Review of optimization of power dispatch in renewable energy system. *Proceedings of the IEEE Innovative Smart Grid Technologies* (IEEE, Melbourne, 2016), pp. 250–257
65. M. Fathi, H. Bevrani, Regulating power management in interconnected microgrids. J. Renew. Sustain. Energy **9**(5), 055502 (2017)
66. M. Fathi, A spectrum allocation scheme between smart grid communication and neighbor communication networks. IEEE Syst. J. **12**(1), 465–472 (2018)
67. P. Hajimirzaee, M. Fathi, N.N. Qader, Quality of service aware traffic scheduling in wireless smart grid communication. Telecommun. Syst. **66**(2), 233–242 (2017)
68. H. Bevrani, *Robust Power System Frequency Control* (Springer, Switzerland, 2014)
69. H. Bevrani, T. Hiyama, *Intelligent Automatic Generation Control* (CRC Press, New York, 2011)
70. H. Bevrani, B. Francois, T. Ise, *Microgrid Dynamics and Control* (Wiley, Hoboken, 2017)
71. H. Bevrani, S. Shokoohi, An intelligent droop control for simultaneous voltage and frequency regulation in islanded microgrids. IEEE Trans. Smart Grid **4**(3), 1505–1513 (2013)
72. H. Bevrani, M. Watanabe, Y. Mitani, *Power System Monitoring and Control* (IEEE-Wiley Press, Hoboken, 2014)

Linear Algebra Review

2

Abstract

Linear algebra is a broad topic in mathematics with a wide range of applications in engineering. In particular, it is the main prerequisite of optimization. To satisfy this requirement, this chapter provides an intensive review of linear algebra with the aim of covering mathematical fundamentals in the next chapters. The chapter begins with an introduction to vector and matrix spaces. It then employs this introduction to analyze the solution of a set of linear equations which often appear in constraints of optimization problems. The chapter then discusses eigenvalues, eigenvectors, matrix diagonalization, and quadratic forms and introduces positive and negative definite properties accordingly. Some calculus concepts such as level set, gradient, Hessian, and directional derivative with a broad range of applications in optimization are finally presented.

Keywords

Directional derivative · Gradient · Hessian · Level set · Matrix diagonalization · Vector space

2.1 Vector Space

Vector as a mathematical concept is to represent a set of numbers in a single array in the form of a column or a row. Here, in this book, vectors are denoted by lowercase boldface letters in a column array. A real vector **a**, consisting of a set of real numbers, is indicated in the form of

$$\mathbf{a} = \begin{bmatrix} a_1 \\ a_2 \\ \vdots \\ a_n \end{bmatrix} \in \mathbb{R}^n \tag{2.1}$$

where $a_1, a_2, \ldots, a_n \in \mathbb{R}$ are entries of **a**. Associated with each vector, there is a transpose vector. The transpose of a column vector **a** is a row vector indicated by $\mathbf{a}^T = [a_1, \ldots, a_n]$ and vice versa. Following is an example in MATLAB:

M. Fathi, H. Bevrani, *Optimization in Electrical Engineering*,
https://doi.org/10.1007/978-3-030-05309-3_2

```
>> a= [4; 5; 1]

a =

     4
     5
     1

>> a'

ans =

     4     5     1
```

where $\mathbf{a}'$ denotes $\mathbf{a}^T$ in MATLAB. Addition and subtraction of any two same size vectors $\mathbf{a}$ and $\mathbf{b}$ is denoted by $\mathbf{a} \pm \mathbf{b} = \begin{bmatrix} a_1 \pm b_1 \\ \vdots \\ a_n \pm b_n \end{bmatrix}$. Moreover, any vector can be scaled using its multiplication by a scalar $\alpha \in \mathbb{R}$ as $\alpha\mathbf{a} = \begin{bmatrix} \alpha a_1 \\ \vdots \\ \alpha a_n \end{bmatrix}$. The real vector space $\mathbb{R}^n$ is closed under vector addition and scalar multiplication. In other words, the results of these operations on vectors in $\mathbb{R}^n$ belong to $\mathbb{R}^n$.

Definition 1 A set of vectors $\{\mathbf{a}_1, \mathbf{a}_2 \ldots, \mathbf{a}_n\}$ is called *linearly dependent* if there exist scalars $\alpha_1, \alpha_2, \ldots, \alpha_n$, not all zeros, such that:

$$\alpha_1\mathbf{a}_1 + \alpha_2\mathbf{a}_2 + \cdots + \alpha_n\mathbf{a}_n = 0. \tag{2.2}$$

In other words, there is at least one scalar $\alpha_i \neq 0$. Alternatively, a set of vectors is said to be linearly independent if the equality in (2.2) implies that all α_i's are zero.

Example 1 Are vectors $\mathbf{a}_1 = \begin{bmatrix} 1 \\ 2 \\ 0 \end{bmatrix}$ and $\mathbf{a}_2 = \begin{bmatrix} 0 \\ 5 \\ 6 \end{bmatrix}$ linearly independent?

Solution The equality $\alpha_1\mathbf{a}_1 + \alpha_2\mathbf{a}_2 = 0$ implies $\alpha_1 = \alpha_2 = 0$. Therefore, they are linearly independent.

Alternatively, if we consider the set $\{\mathbf{a}_1, \mathbf{a}_2, \mathbf{a}_3\}$, where $\mathbf{a}_3 = \begin{bmatrix} 2 \\ 9 \\ 6 \end{bmatrix}$, it can be written that $2\mathbf{a}_1 + \mathbf{a}_2 - \mathbf{a}_3 = 0$. Therefore, the set is linearly dependent.

```
>> a1=[1; 2; 0];
>> a2=[0; 5; 6];
```

```
>> a3 = 2*a1 + a2
a3 =
      2
      9
      6
```

□

As a special case, a set composed of a zero vector $\mathbf{0}$ is linearly dependent. Note that in a zero vector, all entries are zero. Moreover, a set composed of only one vector that is nonzero is linearly independent.

Following in Theorem 1, linearly dependence is stated in terms of linear combination.

Theorem 1 *In a set of vectors* $\{\mathbf{a}_1, \mathbf{a}_2, \ldots, \mathbf{a}_n\}$, *each vector can be written as a linear relation or linear combination of the remaining vectors if and only if the set is linearly dependent.*

Proof If $\{\mathbf{a}_1, \mathbf{a}_2, \ldots, \mathbf{a}_n\}$ is linearly dependent, then there is at least a scalar $\alpha_i \neq 0$ such that $\alpha_1\mathbf{a}_1 + \cdots + \alpha_i\mathbf{a}_i + \cdots + \alpha_n\mathbf{a}_n = 0$. This results in:

$$\mathbf{a}_i = -\frac{\alpha_1}{\alpha_i}\mathbf{a}_1 - \cdots - \frac{\alpha_n}{\alpha_i}\mathbf{a}_n.$$

In other words, $\mathbf{a}_i$ is a linear combination of the other vectors in the given set. Conversely, without loss of generality, suppose that $\mathbf{a}_i$ is a linear combination of the other vectors in the set, i.e., $\mathbf{a}_i = \beta_1\mathbf{a}_1 + \cdots + \beta_n\mathbf{a}_n$, then

$$\beta_1\mathbf{a}_1 + \cdots + (-1)\mathbf{a}_i + \cdots + \beta_n\mathbf{a}_n = 0.$$

Because of nonzero coefficient (-1), the set of vector is thus linearly dependent. □

In Example 1, there is a linear combination between $\mathbf{a}_1$, $\mathbf{a}_2$, and $\mathbf{a}_3$, i.e., $\mathbf{a}_3 = 2\mathbf{a}_1 + \mathbf{a}_2$. Therefore, they are linearly dependent.

To get more insight into the vector space, in the following, a number of well-known terms including subspace, span, basis, and dimension are introduced.

Definition 2 A subset $\mathcal{S}$ of $\mathbb{R}^n$ is called a subspace if $\mathcal{S}$ is closed under vector addition and scalar multiplication. It is apparent that every subspace contains the zero vector $\mathbf{0}$.

Definition 3 The space of all linear combinations of a set of vectors $\{\mathbf{a}_1, \mathbf{a}_2, \ldots, \mathbf{a}_n\}$ is called the span of these vectors. This is written as:

$$\mathsf{span}[\mathbf{a}_1, \mathbf{a}_2, \ldots, \mathbf{a}_n] = \left\{\sum_{i=1}^{n} \alpha_i\mathbf{a}_i | \alpha_i \in \mathbb{R}\right\} \tag{2.3}$$

Definition 4 A set of linearly independent vectors $\{\mathbf{a}_1, \mathbf{a}_2, \ldots, \mathbf{a}_n\} \subset \mathcal{S}$ which spans $\mathcal{S}$, i.e., $\mathcal{S} = \text{span}[\mathbf{a}_1, \mathbf{a}_2, \ldots, \mathbf{a}_n]$, is called a basis for $\mathcal{S}$.

Definition 5 All bases for a subspace $\mathcal{S}$ contain the same number of vectors. This number is called the dimension of $\mathcal{S}$ and is denoted by $\dim \mathcal{S}$.

For example, consider the vector set:

$$\left\{ e_1 = \begin{bmatrix} 1 \\ 0 \\ 0 \end{bmatrix}, e_2 = \begin{bmatrix} 0 \\ 1 \\ 0 \end{bmatrix}, e_3 = \begin{bmatrix} 0 \\ 0 \\ 1 \end{bmatrix} \right\}.$$

These vectors span the real space $\mathbb{R}^3$. Moreover, as e_1, e_2, and e_3 are linearly independent, they are also a basis for $\mathbb{R}^3$, and the dimension of this space is 3. As another example, consider $e_4 = \begin{bmatrix} 1 \\ 1 \\ 0 \end{bmatrix}$ and vector set $\{e_1, e_2, e_3, e_4\}$. This vector set also spans $\mathbb{R}^3$, but is not basis as it is not linearly independent.

The importance of a basis for a subspace relies on the property presented in Theorem 2, which states that any vector in the subspace can be represented as a unique linear combination of a basis.

Theorem 2 *Let* $\{\mathbf{a}_1, \mathbf{a}_2, \ldots, \mathbf{a}_n\}$ *be a basis of subspace* $\mathcal{S}$. *Then, any vector* $\mathbf{a}$ *in* $\mathcal{S}$ *can be represented uniquely as:*

$$\mathbf{a} = \sum_{i=1}^{n} \alpha_i \mathbf{a}_i : \alpha_i \in \mathbb{R}.$$

Proof Assume that there is a vector $\mathbf{a} \in \mathcal{S}$ with two representations:

$$\mathbf{a} = \alpha_1 \mathbf{a}_1 + \cdots + \alpha_n \mathbf{a}_n$$

and

$$\mathbf{a} = \beta_1 \mathbf{a}_1 + \cdots + \beta_n \mathbf{a}_n.$$

These two representations imply $(\alpha_1 - \beta_1)\mathbf{a}_1 + \cdots + (\alpha_n - \beta_n)\mathbf{a}_n = 0$. Because of linearly independent vectors in $\{\mathbf{a}_1, \ldots, \mathbf{a}_n\}$, we can write $\alpha_1 - \beta_1 = \cdots = \alpha_n - \beta_n = 0$ or equivalently $\alpha_1 = \beta_1, \alpha_2 = \beta_2, \ldots, \alpha_n = \beta_n$. Therefore, the linear presentation of $\mathbf{a}$ in terms of the basis $\{\mathbf{a}_1, \mathbf{a}_2, \ldots, \mathbf{a}_n\}$ is unique. □

2.2 Matrix

Matrix is the generalization of the vector concept to represent a set of numbers in two or even more dimensions. Matrices are denoted by capital boldface letters. In the case of two-dimensional matrix, a matrix $\mathbf{A} = \{a_{ij}\} \in \mathbb{R}^{m\times n}$ is a rectangular array of numbers or functions in m rows and n columns:

$$\mathbf{A} = \begin{bmatrix} a_{11} & a_{12} & \dots & a_{1n} \\ a_{21} & a_{22} & \vdots & a_{2n} \\ \vdots & \dots & \ddots & \dots \\ a_{m1} & a_{m2} & \vdots & a_{mn} \end{bmatrix} \tag{2.4}$$

where a_{ij} is the matrix entry of row i and column j. To ease the representation, let us denote the kth column of $\mathbf{A}$ by $\mathbf{a}_k \triangleq \begin{bmatrix} a_{1k} \\ a_{2k} \\ \vdots \\ a_{mk} \end{bmatrix}$. Following this notation, then $\mathbf{A}$ can be written as $\mathbf{A} \triangleq [\mathbf{a}_1, \mathbf{a}_2, \dots, \mathbf{a}_n]$.

Note that two matrices are equal if and only if they have the same number of rows and columns and corresponding entries are equal. Similar to the vector concept, associated with each matrix, there is a transpose matrix. Transpose of $\mathbf{A} \in \mathbb{R}^{m\times n}$ is the substitution of row and column entries and is denoted by $\mathbf{A}^\mathbf{T} \in \mathbb{R}^{n\times m}$ as:

$$\mathbf{A}^T = \begin{bmatrix} \mathbf{a}_1^T \\ \mathbf{a}_2^T \\ \vdots \\ \mathbf{a}_n^T \end{bmatrix} \tag{2.5}$$

where the row vector $\mathbf{a}_j^T$ is the transpose of the column vector $\mathbf{a}_j$. Following are a number of transposition rules:

$$(\mathbf{A}^T)^T = \mathbf{A} \tag{2.6}$$

$$(\mathbf{A} + \mathbf{B})^T = \mathbf{A}^T + \mathbf{B}^T \tag{2.7}$$

$$(c\mathbf{A})^T = c\mathbf{A}^T \tag{2.8}$$

$$(\mathbf{AB})^T = \mathbf{B}^T\mathbf{A}^T \tag{2.9}$$

Matrices appear in any general form. However, the following special matrices are important in understanding the properties of matrices.

- *Square matrix*: a matrix with the same number n of rows and columns is called a square matrix and is denoted by $\mathbf{A} \in \mathbb{R}^{n\times n}$.
- *Symmetric matrix*: a square matrix $\mathbf{A} = \{a_{ij}\} \in \mathbb{R}^{n\times n}$ is called symmetric if $\mathbf{A}^T = \mathbf{A}$, i.e., $a_{ij} = a_{ji}$.

- *Diagonal matrix*: a square matrix $\mathbf{D} = \{d_{ij}\} \in \mathbb{R}^{n\times n}$ that has nonzero entries only on the main diagonal is called diagonal matrix. Entries above and below the main diagonal must be zero, i.e., $d_{ij} = 0$ for $i \neq j$.
- *Identity matrix*: a diagonal matrix $\mathbf{I} = \{t_{ij}\}$ whose all entries in the main diagonal are 1 is called identity or unit matrix, i.e., $t_{ij} = 1$ for $i = j$ and $t_{ij} = 0$ for $i \neq j$. It is noteworthy that for each square matrix $\mathbf{A} \in \mathbb{R}^{n\times n}$ and $\mathbf{I} \in \mathbb{R}^{n\times n}$, $\mathbf{AI} = \mathbf{IA} = \mathbf{A}$.

The matrix concept has a number of useful characteristics, which can be used in matrix manipulation. In the following subsections, four more important characteristics including range and null spaces, rank, determinant, and minor are introduced.

2.2.1 Range and Null Spaces

Associated with each matrix **A**, range and null spaces are stated in Definitions 6 and 7.

Definition 6 Let $\mathbf{a}_1, \mathbf{a}_2, \ldots, \mathbf{a}_n$ be columns of $\mathbf{A} \in \mathbb{R}^{m\times n}$. Then, the range space of **A** is defined as:

$$\mathcal{R}(\mathbf{A}) \triangleq \{\mathbf{Ax} = x_1\mathbf{a}_1 + x_2\mathbf{a}_2 + \cdots + x_n\mathbf{a}_n | \mathbf{x} \in \mathbb{R}^n\} \tag{2.10}$$

or equivalently $\mathcal{R}(\mathbf{A}) \triangleq \text{span}[\mathbf{a}_1, \mathbf{a}_2, \ldots, \mathbf{a}_n]$.

Definition 7 The null space of $\mathbf{A} \in \mathbb{R}^{m\times n}$ is defined as:

$$\mathcal{N}(\mathbf{A}) \triangleq \{\mathbf{x} \in \mathbb{R}^n | \mathbf{Ax} = \mathbf{0}\}. \tag{2.11}$$

Moreover, the dimension of $\mathcal{N}(\mathbf{A})$ is called nullity.

Definition 8 Row operations consist of interchanging of two rows, multiplying a row by a nonzero scalar, and replacing a row with the combination of itself plus a multiple of another row. It is noteworthy that under row operations, the rank, range space, and null space of a matrix will not change. Therefore, row operations can be used to find these characteristics.

Example 2 Find null space of $A = \begin{bmatrix} 4 & -2 & 0 \\ 2 & 1 & -1 \\ 2 & -3 & 1 \end{bmatrix}$.

Solution Each row of **A** represents coefficients of corresponding equation in $\mathbf{Ax} = \mathbf{0}$. Therefore, using row operations on **A**, we derive

$$\begin{bmatrix} 4 & -2 & 0 \\ 2 & 1 & -1 \\ 2 & -3 & 1 \end{bmatrix} \xrightarrow[r_3 \triangleq r_3 - r_1/2]{r_2 \triangleq r_2 - r_1/2} \begin{bmatrix} 4 & -2 & 0 \\ 0 & 2 & -1 \\ 0 & -2 & 1 \end{bmatrix} \xrightarrow{r_3 \triangleq r_3 + r_2} \begin{bmatrix} 4 & -2 & 0 \\ 0 & 2 & -1 \\ 0 & 0 & 0 \end{bmatrix}$$

Therefore, the set of equations $\mathbf{Ax} = 0$ reduces to $\Rightarrow \begin{cases} 4x_1 - 2x_2 = 0 \\ 2x_2 - x_3 = 0 \end{cases}$ that results in $x_2 = \frac{1}{2}x_3$, $x_1 = \frac{1}{2}x_2 = \frac{1}{4}x_3$. Assuming any value for x_3, the solution is in the form of:

$$\mathbf{x} = \begin{bmatrix} x_1 \\ x_2 \\ x_3 \end{bmatrix} = \begin{bmatrix} 1/4 \\ 1/2 \\ 1 \end{bmatrix} x_3$$

and therefore the null space can be represented by:

$$\mathcal{N}(\mathbf{A}) = \left\{ \begin{bmatrix} 1/4 \\ 1/2 \\ 1 \end{bmatrix} c \mid c \in \mathbb{R} \right\}$$

with dimension or nullity 1. This can be also derived in MATLAB using:

```
>> A = [4 -2 0; 2 1 -1; 2 -3 1];
>> null(A)

ans =

    0.2182
    0.4364
    0.8729
```

Note that MATLAB solution has been normalized to be of length unity. □

2.2.2 Rank

The maximum number of linearly independent columns of matrix **A** is called rank of **A** and is denoted by rank **A**. In other words, rank is the dimension of span$[\mathbf{a}_1, \mathbf{a}_2, \ldots, \mathbf{a}_n]$. The rank of a matrix is not changed under the following operations:

- Multiplication of columns by nonzero scalars.
- Interchange of columns.
- Addition to a given column a linear combination of the other columns.

These operations can be used to determine the rank of a given matrix, as shown in Example 3.

Example 3 Find rank of $\mathbf{A}=\begin{bmatrix} 3 & 0 & 2 & 2 \\ -6 & 42 & 24 & 54 \\ 21 & -21 & 0 & -15 \end{bmatrix}$.

Solution Let $\mathbf{a}_i$ be the ith column of $\mathbf{A}$. Using the aforementioned column operations, we derive

$$\begin{bmatrix} 3 & 0 & 2 & 2 \\ -6 & 42 & 24 & 54 \\ 21 & -21 & 0 & -15 \end{bmatrix} \xrightarrow{\mathbf{a}_1 \triangleq \mathbf{a}_1+\mathbf{a}_2} \begin{bmatrix} 3 & 0 & 2 & 2 \\ 36 & 42 & 24 & 54 \\ 0 & -21 & 0 & -15 \end{bmatrix} \xrightarrow{\mathbf{a}_1 \triangleq \mathbf{a}_1-3/2\mathbf{a}_3} \begin{bmatrix} 0 & 0 & 2 & 2 \\ 0 & 42 & 24 & 54 \\ 0 & -21 & 0 & -15 \end{bmatrix} \xrightarrow{\mathbf{a}_4 \triangleq \mathbf{a}_4-(5/7\mathbf{a}_2+\mathbf{a}_3)} \begin{bmatrix} 0 & 0 & 2 & 0 \\ 0 & 42 & 24 & 0 \\ 0 & -21 & 0 & 0 \end{bmatrix}$$

$\Rightarrow$ rank $\mathbf{A} = 2$ due to the two remained nonzero linearly independent vectors. This result can also be achieved in MATLAB using:

```
>> A = [3 0 2 2; −6 42 24 54; 21 −21 0 −15];
>> rank(A)

ans =
     2
```

□

For each matrix $\mathbf{A} \in \mathbb{R}^{m\times n}$, it can be shown that rank $\mathbf{A} \le \min(m, n)$.

Theorem 3 *For each matrix* **A***, rank* **A** *= Number of columns−nullity.*

Instead of providing a proof, this theorem is verified for Example 3, where the number of columns is 4 and the rank is 2. The nullity is derived in MATLAB using:

```
>> A = [3 0 2 2; −6 42 24 54; 21 −21 0 −15];
>> rank(null(A))

ans =

     2
```

which confirms the statement in Theorem 3.

2.2.3 Determinant

Determinant is an operation defined for square matrices. The determinant of a square matrix $\mathbf{A} \in \mathbb{R}^{n\times n}$ is a scalar function det: $\mathbb{R}^{n\times n} \to \mathbb{R}$ denoted by $\det \mathbf{A}$ or $|\mathbf{A}|$. If $\det \mathbf{A} \neq 0$, $\mathbf{A}$ is said to be nonsingular matrix, otherwise it is a singular matrix. The determinant function has a number of useful properties, which can be used in matrix manipulation. Before going into these properties, let's begin with some simple order determinants.

- First-order determinant: if $\mathbf{A} = a \Rightarrow \det \mathbf{A} = a$.
- Second-order determinant: if $\mathbf{A} = \begin{bmatrix} a_{11} & a_{12} \\ a_{21} & a_{22} \end{bmatrix} \Rightarrow \det \mathbf{A} = \begin{vmatrix} a_{11} & a_{12} \\ a_{21} & a_{22} \end{vmatrix} = a_{11}a_{22} - a_{12}a_{21}$.
- Third-order determinant: if $\mathbf{A} = \begin{bmatrix} a_{11} & a_{12} & a_{13} \\ a_{21} & a_{22} & a_{23} \\ a_{31} & a_{32} & a_{33} \end{bmatrix} \Rightarrow \det \mathbf{A} = \begin{vmatrix} a_{11} & a_{12} & a_{13} \\ a_{21} & a_{22} & a_{23} \\ a_{31} & a_{32} & a_{33} \end{vmatrix} = a_{11}\begin{vmatrix} a_{22} & a_{23} \\ a_{32} & a_{33} \end{vmatrix} - a_{21}\begin{vmatrix} a_{12} & a_{13} \\ a_{32} & a_{33} \end{vmatrix} + a_{31}\begin{vmatrix} a_{12} & a_{13} \\ a_{22} & a_{23} \end{vmatrix}$. As shown, the third-order determinant has been reduced to the second-order determinant. Similarly, higher-order determinants can also be reduced to lower-order ones.

Example 4 Find determinant of $\mathbf{A} = \begin{bmatrix} 5 & 4 & 3 \\ 0 & -1 & 7 \\ 0 & 0 & 2 \end{bmatrix}$.

Solution $\det \mathbf{A} = 5\begin{vmatrix} -1 & 7 \\ 0 & 2 \end{vmatrix} = (5)(-1)(2) = -10$, i.e., the product of diagonal entries. This can also be verified in MATLAB using:

```
>> A = [5 4 3; 0 -1 7; 0 0 2];
>> det(A)


ans =

   -10
```

□

Two important properties of the determinant function are stated in Theorems 4 and 5.

Theorem 4 *Determinant of a matrix is a linear function of each column. Consider the kth column of* $\mathbf{A} = [\mathbf{a}_1, \ldots, \mathbf{a}_n]$ *as* $\mathbf{a}_k = \alpha\mathbf{a}_k^{(1)} + \beta\mathbf{a}_k^{(2)}$, *where* $\alpha, \beta \in \mathbb{R}$. *Then:*

$$\begin{aligned} det[\mathbf{a}_1, \ldots, \alpha\mathbf{a}_k^{(1)} + \beta\mathbf{a}_k^{(2)}, \ldots, \mathbf{a}_n] &= \alpha\, det[\mathbf{a}_1, \ldots, \mathbf{a}_k^{(1)}, \ldots, \mathbf{a}_n] \\ &+ \beta\, det[\mathbf{a}_1, \ldots, \mathbf{a}_k^{(2)}, \ldots, \mathbf{a}_n]. \end{aligned} \tag{2.12}$$

Theorem 5 *Determinant of a matrix with two identical columns or rows is zero. In other words, if for some i and j, we have $\mathbf{a}_i = \mathbf{a}_j$, then* $\det \mathbf{A} = 0$.

The proof of Theorems 4 and 5 can be found in the literature of linear algebra [1,2]. Considering these theorems, the following results can be derived.

1. Determinant of a matrix with a zero column is zero, i.e., $\det[\mathbf{a}_1, \ldots, 0, \ldots, \mathbf{a}_n] = 0$. This is derived from Theorem 4 when $\alpha = \beta = 0$.
2. Determinant does not change if we add to a column a linear combination of the other columns. This implies that columns of $\mathbf{A}$ are linearly dependent if and only if $\det \mathbf{A} = 0$.
3. Determinant changes its sign if we interchange columns, that is:

$$\det[\mathbf{a}_1, \ldots, \mathbf{a}_i, \ldots, \mathbf{a}_j, \ldots \mathbf{a}_n] = -\det[\mathbf{a}_1, \ldots, \mathbf{a}_j, \ldots, \mathbf{a}_i, \ldots \mathbf{a}_n].$$

4. Let $\mathbf{I}_n \in \mathbb{R}^{n \times n}$ be the identity matrix, then $\det \mathbf{I} = 1$.

2.2.4 Minor

A pth-order *minor* of an $m \times n$ matrix $\mathbf{A}$ with $p \leq \min(m, n)$ is the determinant of a $p \times p$ matrix obtained by deleting $m - p$ rows and $n - p$ columns. Therefore, if there is a nonzero minor, the columns associated with this nonzero minor are linearly independent. Consequently, the rank of a matrix is equal to the highest order of its nonzero minors.

2.3 Linear Equations

Linear equations are often raised in the constraints of optimization problems. Therefore, it is of high importance to investigate the quantity of feasible solutions for linear equations. Consider m linear equations in n unknowns as:

$$\begin{aligned} a_{11}x_1 + a_{12}x_2 + \cdots + a_{1n}x_n &= b_1 \\ a_{21}x_1 + a_{22}x_2 + \cdots + a_{2n}x_n &= b_2 \\ &\vdots \\ a_{m1}x_1 + a_{m2}x_2 + \cdots + a_{mn}x_n &= b_m \end{aligned} \tag{2.13}$$

where x_i's are unknowns and a_{ij}'s are coefficients. This set of equations can be represented as $x_1\mathbf{a}_1 + x_2\mathbf{a}_2 + \cdots + x_n\mathbf{a}_n = \mathbf{b}$, where $\mathbf{a}_j \triangleq \begin{bmatrix} a_{1j} \\ \vdots \\ a_{mj} \end{bmatrix}$ and $\mathbf{b} \triangleq \begin{bmatrix} b_1 \\ \vdots \\ b_m \end{bmatrix}$. Alternatively, if $\mathbf{A} \triangleq [\mathbf{a}_1, \mathbf{a}_2, \ldots, \mathbf{a}_n]$, the set of equations in (2.13) is represented as:

$$\mathbf{A}\mathbf{x} = \mathbf{b} \tag{2.14}$$

where $\mathbf{x} \triangleq \begin{bmatrix} x_1 \\ \vdots \\ x_n \end{bmatrix}$ is the vector of unknowns and each row of $\mathbf{A}$ represents the coefficients of an equivalent equation.

The discussions on the possible solutions of a set of linear equations are summarized in Theorem 6 [1]. The proof is left as an exercise to the reader.

Theorem 6 *Let's include the vector* $\mathbf{b} \in \mathbb{R}^m$ *as a column in* $\mathbf{A} \in \mathbb{R}^{m \times n}$ *and denote the resulted matrix as* $[\mathbf{A}\ \mathbf{b}]$, *usually known as the augmented matrix. Then, the system of linear equation* $\mathbf{Ax} = \mathbf{b}, \mathbf{x} \in \mathbb{R}^n$ *has:*

- *A unique solution if and only if* $rank\ \mathbf{A} = rank\ [\mathbf{A}\ \mathbf{b}] = n$
- *Infinite number of solutions if and only if* $rank\ \mathbf{A} = rank\ [\mathbf{A}\ \mathbf{b}] < n$
- *No solution if and only if* $rank\ \mathbf{A} < rank\ [\mathbf{A}\ \mathbf{b}]$.

Example 5 Find the number of solutions in the following set of equations:

$$x_1 - x_2 + x_3 = 0$$

$$-x_1 + x_2 - x_3 = 0$$

$$10x_2 + 25x_3 = 90$$

$$20x_1 + 10x_2 = 80$$

Solution The augmented matrix is $[\mathbf{A}\ \mathbf{b}] = \begin{bmatrix} 1 & -1 & 1 & 0 \\ -1 & 1 & -1 & 0 \\ 0 & 10 & 25 & 90 \\ 20 & 10 & 0 & 80 \end{bmatrix}$. Using column operations as described in Example 3, we derive rank $[\mathbf{A}\ \mathbf{b}] = \text{rank}\ \mathbf{A} = 3$. Considering Theorem 6, this equation set has a unique solution. It can be computed in MATLAB as:

```
>> A = [1 −1 1; −1 1 −1; 0 10 25; 20 10 0];
>> b = [0; 0; 90; 80];
>> x = A\b

x =

    2.0000
    4.0000
    2.0000
```

□

2.4 Inverse

The inverse of square matrix $\mathbf{A} \in \mathbb{R}^{n\times n}$ is denoted by $\mathbf{A}^{-1} \in \mathbb{R}^{n\times n}$ such that:

$$\mathbf{A}\mathbf{A}^{-1} = \mathbf{A}^{-1}\mathbf{A} = \mathbf{I} \tag{2.15}$$

where $\mathbf{I}$ is the identity matrix. The inverse of a matrix does not exist in general. If $\mathbf{A}$ has an inverse, then $\mathbf{A}$ is called a nonsingular matrix. Otherwise, it is called a singular matrix.

Example 6 Find inverse of $\mathbf{A} = \begin{bmatrix} 5 & 4 & 3 \\ 0 & -1 & 7 \\ 0 & 0 & 2 \end{bmatrix}$.

Solution This can be done in MATLAB using:

```
>> A = [5 4 3; 0 -1 7; 0 0 2];
>> inv(A)

ans =

    0.2000    0.8000   -3.1000
         0   -1.0000    3.5000
         0         0    0.5000

>> A*inv(A)

ans =

    1.0000         0   -0.0000
         0    1.0000         0
         0         0    1.0000
```

□

The condition on existing of the inverse of a square matrix is stated in Theorem 7.

Theorem 7 *The inverse* $\mathbf{A}^{-1}$ *of* $\mathbf{A} \in \mathbb{R}^{n\times n}$ *exists if and only if rank* $\mathbf{A} = n$*, i.e.,* $\mathbf{A}$ *is full rank.*

Proof Let $\mathbf{b} \in \mathbb{R}^n$ be a linear combination of columns of $\mathbf{A}$. To prove the theorem, first assume that $\mathbf{A}^{-1}$ exists. Multiplying both sides of $\mathbf{Ax} = \mathbf{b}$ by $\mathbf{A}^{-1}$, we derive $\mathbf{x} = \mathbf{A}^{-1}\mathbf{b}$ that is a unique solution. Following Theorem 6, $\mathbf{A}$ must have rank n.
Conversely, let rank $\mathbf{A} = n$. Then, the system of equation $\mathbf{Ax} = \mathbf{b}$ has a unique solution in the form of $\mathbf{x} = \mathbf{Bb}$, where $\mathbf{B}$ is an $n \times n$ matrix. Substituting $\mathbf{x}$ in $\mathbf{Ax} = \mathbf{b} \Rightarrow \mathbf{A}(\mathbf{Bb}) = \mathbf{b} \Rightarrow (\mathbf{AB})\mathbf{b} =$

$\mathbf{b} \Rightarrow \mathbf{AB} = \mathbf{I}$. On the other hand, $\mathbf{x} = \mathbf{Bb} = \mathbf{BAx} \Rightarrow \mathbf{BA} = \mathbf{I}$. From both $\mathbf{AB} = \mathbf{I}$ and $\mathbf{BA} = \mathbf{I}$, it is concluded that $\mathbf{B} = \mathbf{A}^{-1}$. □

Theorem 8 *If* $\mathbf{A}$ *has an inverse, the inverse is unique.*

Proof Assume that two matrices $\mathbf{B}$ and $\mathbf{C}$ are inverses of $\mathbf{A}$. Using $\mathbf{B} = \mathbf{BI} = \mathbf{B}(\mathbf{AC}) = (\mathbf{BA})\mathbf{C} = \mathbf{IC} = \mathbf{C}$, it is concluded that $\mathbf{B} = \mathbf{C}$. Therefore, the inverse is unique. □

2.5 Inner Product

For every $\mathbf{a}$ and $\mathbf{b} \in \mathbb{R}^n$, the inner product function $< .,. >: \mathbb{R}^n \times \mathbb{R}^n \to \mathbb{R}$ is defined as:

$$\langle \mathbf{a}, \mathbf{b} \rangle = \mathbf{a}^T\mathbf{b} = \sum_{i=1}^{n} a_i b_i = a_1 b_1 + \cdots + a_n b_n. \tag{2.16}$$

For example, if:

$$\mathbf{a} = \begin{bmatrix} 1 \\ 5 \\ 7 \end{bmatrix}, \mathbf{b} = \begin{bmatrix} 3 \\ 4 \\ 1 \end{bmatrix} \Rightarrow \langle \mathbf{a}, \mathbf{b} \rangle = (1)(3) + (5)(4) + (7)(1) = 30.$$

```
>> a=[1; 5; 7];
>> b=[3; 4; 1];
>> a'*b

ans =

    30
```

In particular, the vectors $\mathbf{a}$ and $\mathbf{b}$ are said to be orthogonal if $\langle \mathbf{a}, \mathbf{b} \rangle = 0$. Moreover, $\langle \mathbf{a}, \mathbf{a} \rangle \geq 0$ and $\langle \mathbf{a}, \mathbf{a} \rangle = 0$ if and only if $\mathbf{a} = \mathbf{0}$.

A well-known inequality in the context of inner product is Cauchy–Schwarz inequality stated in Theorem 9. The proof can be found in the literature of linear algebra [1,2].

Theorem 9 (Cauchy–Schwarz Inequality) *For any two vectors* $\mathbf{a}$, $\mathbf{b} \in R^n$, *the inequality:*

$$|\langle \mathbf{a}, \mathbf{b} \rangle|^2 \leq \langle \mathbf{a}, \mathbf{a} \rangle \langle \mathbf{b}, \mathbf{b} \rangle$$

holds. The equality holds if $\mathbf{a} = \alpha\mathbf{b}$ *for* $\alpha \in \mathbb{R}$.

2.6 Norm

For any $\mathbf{a} \in \mathbb{R}^n$, the pth-order norm function is denoted by $\|\mathbf{a}\|_p$. The mathematical operation depends on p. Following are a number of well-known norm operations:

- 2-norm or Euclidean distance : $\|\mathbf{a}\|_2 = \sqrt{\langle \mathbf{a}, \mathbf{a} \rangle} = \sqrt{\mathbf{a}^T \mathbf{a}} = \sqrt{a_1^2 + \cdots + a_n^2}$.
- 1-norm: $\|\mathbf{a}\|_1 = |a_1| + |a_2| + \cdots + |a_n|$.
- ∞-norm: $\|\mathbf{a}\|_\infty = \max\{|a_1|, |a_2|, \ldots, |a_n|\}$.
- p-norm: $\|\mathbf{a}\|_p = (|a_1|^p + |a_2|^p + \cdots + |a_n|^p)^{1/p}$.

Any pth-order norm operation of **a** can be computed in MATLAB using `norm(a,p)` command. Norm operations have the following properties:

- $\|\mathbf{a}\| \geq 0$, $\|\mathbf{a}\| = 0$ if and only if $\mathbf{a} = \mathbf{0}$.
- $\|r\mathbf{a}\| = |r|\|\mathbf{a}\|$, $r \in \mathbb{R}$.
- Triangular inequality: $\|\mathbf{a} + \mathbf{b}\| \leq \|\mathbf{a}\| + \|\mathbf{b}\|$ (this can be concluded from Cauchy–Schwarz inequality).
- In the usual 2-dimensional space, $\langle \mathbf{a}, \mathbf{b} \rangle = \|\mathbf{a}\| \|\mathbf{b}\| \cos\theta$, where θ is the angle between **a** and **b**. This is a geometric representation of the inner product.

Finally, the norm operation is also extended for matrices. In particular, the *Frobenius* norm of matrix $\mathbf{A} \in \mathbb{R}^{m\times n}$ is defined as:

$$\|\mathbf{A}\|_F = \sqrt{\sum_{i=1}^{m}\sum_{j=1}^{n} a_{ij}^2}. \tag{2.17}$$

2.7 Eigenvalues and Eigenvectors

Consider a square matrix $\mathbf{A} \in \mathbb{R}^{n\times n}$. A scalar λ is called an *eigenvalue* of **A** if there is a vector $\mathbf{v} \neq 0 \in \mathbb{R}^n$ such that:

$$\mathbf{A}\mathbf{v} = \lambda\mathbf{v}. \tag{2.18}$$

v is also called the *eigenvector* corresponding to eigenvalue λ. Equation (2.18) says that **Av** does not change the direction of **v**, it only can change the magnitude and/or sign in the direction of **v** [4].

From $\mathbf{A}\mathbf{v} = \lambda\mathbf{v}$, we derive $(\mathbf{A} - \lambda\mathbf{I})\mathbf{v} = 0$, where **I** is the $n \times n$ identity matrix. Because of nonzero solution **v** for this equation, it is concluded that matrix $(\mathbf{A} - \lambda\mathbf{I})$ is singular and its columns are linearly dependent. Consequently, the necessary and sufficient condition for existing eigenvalue λ is $\det(\mathbf{A} - \lambda\mathbf{I}) = 0$. This alternatively results in a characteristic equation as:

$$\lambda^n + a_{n-1}\lambda^{n-1} + \cdots + a_1\lambda_1 + a_0 = 0 \tag{2.19}$$

with n roots as eigenvalues.

Example 7 Find eigenvalues and eigenvectors of $A = \begin{bmatrix} -5 & 2 \\ 2 & -2 \end{bmatrix}$.

Solution $\det(\mathbf{A} - \lambda I) = \begin{vmatrix} -5-\lambda & 2 \\ 2 & -2-\lambda \end{vmatrix} = (-5-\lambda)(-2-\lambda) - 4 = \lambda^2 + 7\lambda + 6 = 0 \Rightarrow \lambda_1 = -6, \lambda_2 = -1$. A set of feasible solutions for $\mathbf{A}\mathbf{v}_1 = \lambda_1\mathbf{v}_1$ and $\mathbf{A}\mathbf{v}_2 = \lambda_2\mathbf{v}_2$ are $\mathbf{v}_1 = \begin{bmatrix} 1 \\ -0.5 \end{bmatrix}$ and $\mathbf{v}_2 = \begin{bmatrix} 1 \\ 2 \end{bmatrix}$, respectively. Eigenvectors $\mathbf{v}_1$ and $\mathbf{v}_2$ are linearly independent. □

Moreover, the following command in MATLAB returns the normalized eigenvectors in **V** and eigenvalues in the main diagonal of **D**. Note that the eigenvectors have been normalized to be of length unity.

```
>> A = [-5 2; 2 -2];
>> [V,D] = eig(A)

V =

   -0.8944   -0.4472
    0.4472   -0.8944

D =

   -6     0
    0    -1
```

Theorem 10 *Suppose that* $\mathbf{v}_1, \mathbf{v}_2, \ldots, \mathbf{v}_n$ *are eigenvectors corresponding to distinct eigenvalues* $\lambda_1, \lambda_2, \ldots, \lambda_n$*. Then,* $\mathbf{v}_1, \mathbf{v}_2, \ldots, \mathbf{v}_n$ *are linearly independent vectors.*

The proof can be found in the literature of linear algebra [1, 2].

2.8 Matrix Diagonalization

Matrix diagonalization is the process of taking a square matrix $\mathbf{A} \in \mathbb{R}^{n\times n}$ and transforming it into a special form of:

$$\mathbf{A} = \mathbf{V}\mathbf{D}\mathbf{V}^{-1} \tag{2.20}$$

in which **V** and $\mathbf{D} \in \mathbb{R}^{n\times n}$ and more importantly **D** is in a diagonal form. Here, matrix **V** diagonalizes matrix **A**. The process of diagonalization is described in Theorem 11.

Theorem 11 *The square matrix* $\mathbf{A} \in \mathbb{R}^{n\times n}$ *is diagonalizable if and only if it has n linearly independent eigenvectors.*

Proof Assume that **A** has n linearly independent eigenvectors $\{\mathbf{v}_1, \mathbf{v}_2, \ldots, \mathbf{v}_n\}$ and corresponding distinct eigenvalues $\{\lambda_1, \lambda_2, \ldots, \lambda_n\}$, according to Theorem 10. Let $\mathbf{V} \triangleq [\mathbf{v}_1, \mathbf{v}_2, \ldots, \mathbf{v}_n]$ and write

$$\mathbf{AV} = \mathbf{A}[\mathbf{v}_1, \ldots, \mathbf{v}_n] = [\mathbf{Av}_1, \ldots, \mathbf{Av}_n] = [\lambda_1\mathbf{v}_1, \ldots, \lambda_n\mathbf{v}_n] = \mathbf{VD}$$

where:

$$\mathbf{D} \triangleq \begin{bmatrix} \lambda_1 & 0 & \ldots & 0 \\ 0 & \lambda_2 & \vdots & 0 \\ \vdots & \ldots & \ddots & \ldots \\ 0 & 0 & \vdots & \lambda_n \end{bmatrix}$$

is a diagonal matrix containing eigenvalues in the main diagonal. As a result of linearly independent eigenvectors, the matrix $\mathbf{V}$ is invertible and we can write $\mathbf{A} = \mathbf{VDV}^{-1}$ or equivalently $\mathbf{D} = \mathbf{V}^{-1}\mathbf{AV}$.

Conversely, if $\mathbf{A}$ is diagonalizable, then $\mathbf{V}$ the matrix of eigenvectors is invertible. Therefore, eigenvectors are linearly independent. □

Example 8 Find the matrix that diagonalizes $\begin{bmatrix} 2 & 6 \\ 0 & -1 \end{bmatrix}$.

Solution First, we find eigenvalues and eigenvectors:

$$\det(\mathbf{A} - \lambda\mathbf{I}) = \begin{vmatrix} 2-\lambda & 6 \\ 0 & -1-\lambda \end{vmatrix} = (2-\lambda)(-1-\lambda) = 0 \Rightarrow \lambda_1 = 2, \lambda_2 = -1$$

Since eigenvalues are distinct, according to Theorem 11, we claim that the matrix $\mathbf{D} = \begin{bmatrix} 2 & 0 \\ 0 & -1 \end{bmatrix}$ that contains eigenvalues on the main diagonal can diagonalize $\mathbf{A}$. As a verification, note that a set of feasible solutions for $\mathbf{Av}_1 = \lambda_1\mathbf{v}_1$ and $\mathbf{Av}_2 = \lambda_2\mathbf{v}_2$ are $\mathbf{v}_1 = \begin{bmatrix} 1 \\ 0 \end{bmatrix}$ and $\mathbf{v}_2 = \begin{bmatrix} -2 \\ 1 \end{bmatrix}$, respectively. Therefore, we form $\mathbf{V}$ as $\mathbf{V} = \begin{bmatrix} 1 & -2 \\ 0 & 1 \end{bmatrix}$ and moreover $\mathbf{V}^{-1} = \begin{bmatrix} 1 & 2 \\ 0 & 1 \end{bmatrix}$. Substituting in (2.20), we derive $\mathbf{VDV}^{-1} = \begin{bmatrix} 2 & 6 \\ 0 & -1 \end{bmatrix}$ that is the same matrix of $\mathbf{A}$. The solution can also be done in MATLAB as:

```
>> A = [2 6; 0 -1];
>> [V,D] = eig(A)

V =

    1.0000   -0.8944
         0    0.4472

D =

     2     0
     0    -1
```

□

For square matrices that are real and symmetric, the diagonalization derivation can be more simplified. Before doing this, consider Theorem 12.

Theorem 12 *Eigenvectors of real symmetric matrix* $\mathbf{A} \in \mathbb{R}^{n \times n}$ *are mutually orthogonal.*

Proof Here, the theorem is proved only for the case of distinct eigenvalues. Let $\{\lambda_1, \ldots, \lambda_n\}$ be n distinct eigenvalues of $\mathbf{A}$ corresponding to eigenvectors $\{\mathbf{v}_1, \ldots, \mathbf{v}_n\}$. For any λ_i and λ_j, we have

$$\lambda_i \mathbf{v}_i^T \mathbf{v}_j = (\lambda_i \mathbf{v}_i)^T \mathbf{v}_j = (\mathbf{A}\mathbf{v}_i)^T \mathbf{v}_j = \mathbf{v}_i^T \mathbf{A}^T \mathbf{v}_j.$$

Now, $\mathbf{A}^T = \mathbf{A}$ implies $\lambda_i \mathbf{v}_i^T \mathbf{v}_j = \mathbf{v}_i^T \mathbf{A}\mathbf{v}_j = \mathbf{v}_i^T \lambda_j \mathbf{v}_j = \lambda_j \mathbf{v}_i^T \mathbf{v}_j$. If $\lambda_i \neq \lambda_j \Rightarrow \mathbf{v}_i^T \mathbf{v}_j = 0$. Therefore, eigenvalues are mutually orthogonal. □

For square matrices, if each eigenvector $\mathbf{v}_i$ is also normalized to be of length unity, then $\mathbf{v}_i^T \mathbf{v}_i = 1$ for all i. Under the assumption of $\mathbf{V} = [\mathbf{v}_1, \mathbf{v}_2, \ldots, \mathbf{v}_n]$ and considering two derivations $\mathbf{v}_i^T \mathbf{v}_j = 0$ and $\mathbf{v}_i^T \mathbf{v}_i = 1$, we conclude $\mathbf{V}^T \mathbf{V} = \mathbf{I}$ and equivalently to $\mathbf{V}^{-1} = \mathbf{V}^T$. Substituting in (2.20), we finally derive

$$\mathbf{A} = \mathbf{V}\mathbf{D}\mathbf{V}^T = \sum_{i=1}^{n} \lambda_i \mathbf{v}_i \mathbf{v}_i^T \tag{2.21}$$

which is known as the diagonalization for real, square, and symmetric matrix $\mathbf{A}$ [4].

2.9 Singular Value Decomposition

Matrix diagonalization in Sect. 2.8 is valid only for square matrices. For the case of non-square matrices, the diagonalization appears in the form of singular value decomposition that is the topic of this section.

2.9.1 Singular Value

Definition 9 The singular values of an $m \times n$ matrix $\mathbf{A}$ with $m > n$ are the square roots of the nonzero eigenvalues of the symmetric $n \times n$ matrix $\mathbf{A}^T\mathbf{A}$ listed with their multiplicities in decreasing order $\sigma_1 \geq \sigma_2 \geq \cdots \geq \sigma_n \geq 0$. If $m < n$, then the nonzero eigenvalues of the symmetric $m \times m$ matrix $\mathbf{A}\mathbf{A}^T$ are considered rather than $\mathbf{A}^T\mathbf{A}$.

Example 9 Find the singular values for matrix $\mathbf{A} = \begin{bmatrix} 0 & 1 \\ 1 & 1 \\ 1 & 0 \end{bmatrix}$.

Solution Consider $\mathbf{A}^T\mathbf{A} = \begin{bmatrix} 0 & 1 & 1 \\ 1 & 1 & 0 \end{bmatrix} \begin{bmatrix} 0 & 1 \\ 1 & 1 \\ 1 & 0 \end{bmatrix} = \begin{bmatrix} 2 & 1 \\ 1 & 2 \end{bmatrix}$. The characteristic equation is $\det(\mathbf{A}^T\mathbf{A} - \lambda\mathbf{I}_2) = (2-\lambda)^2 - 1 = 0$. Then, we have two eigenvalues $\lambda_1 = 3$ and $\lambda_2 = 1$. According to Definition 9, the corresponding singular values are $\sigma_1 = \sqrt{3}$ and $\sigma_2 = 1$. □

Remark 1 Let $\sigma_{\max}(\mathbf{A})$ be the maximum singular value of an $m \times n$ matrix. Then, 2-norm of $\mathbf{A}$ is defined as $\|\mathbf{A}\|_2 = \sigma_{\max}(\mathbf{A}) = \sqrt{\lambda_{\max}\left(\mathbf{A}^T\mathbf{A}\right)}$.

2.9.2 Singular Value Decomposition

The singular value decomposition (SVD) of a matrix $\mathbf{A}$ is very useful in the context of least squares problems and analyzing properties of a matrix. The existence of decomposition is investigated in Theorem 13.

Theorem 13 *Let* $\mathbf{A}$ *be an* $m \times n$ *matrix with* $m \geq n$*. Then, there exist symmetric orthogonal matrices* $\mathbf{U}_{m\times m}$ *and* $\mathbf{V}_{n\times n}$ *and a block diagonal matrix* $\mathbf{\Sigma}_{m\times n} = diag(\sigma_1, \sigma_2, \ldots, \sigma_n)$ *with order* $\sigma_1 \geq \sigma_2 \geq \cdots \geq \sigma_n \geq 0$ *such that:*

$$\mathbf{A} = \mathbf{U\Sigma V}^T. \tag{2.22}$$

The proof can be found in the literature of linear algebra [2]. The column vectors of $\mathbf{U} = [\mathbf{u}_1, \ldots, \mathbf{u}_m]$ are called the *left singular vectors*, and similarly the columns of $\mathbf{V} = [\mathbf{v}_1, \ldots, \mathbf{v}_n]$ are the *right singular vectors* of matrix $\mathbf{A}$. The columns of $\mathbf{U}$ and $\mathbf{V}$ are orthonormal, and the matrix $\mathbf{\Sigma}$ is diagonal with positive real entries of σ_i. Unlike the more commonly used spectral decomposition in linear algebra, the SVD is defined for all matrices (rectangular or square).

Let $l \triangleq min(m, n)$, then the matrix $\mathbf{\Sigma}$ can be represented as

$$\mathbf{\Sigma} = \begin{bmatrix} \tilde{\mathbf{\Sigma}}_{l\times l} & \mathbf{0}_{l\times(n-l)} \\ \mathbf{0}_{(m-l)\times l} & \mathbf{0}_{(m-l)\times(n-l)} \end{bmatrix} \tag{2.23}$$

where $\tilde{\mathbf{\Sigma}} = diag(\sigma_1, \sigma_2, \ldots, \sigma_l)$ and $\sigma_1 \geq \sigma_2 \geq \ldots, \geq \sigma_l \geq 0$.

Since the matrix $\mathbf{\Sigma}$ must be of dimension $m \times n$ (the same size as $\mathbf{A}$), to maintain dimensional consistency, it should be padded with zeros either on the bottom or to the right of the diagonal block, depending on whether $m > n$ or $m < n$, respectively. In MATLAB, the function `svd` computes the SVD as `[U, S, V] = svd(A)`, where `S` denotes $\mathbf{\Sigma}$.

If $\sigma_r > 0$ is the smallest singular value greater than zero, then the matrix $\mathbf{A}$ has rank r, and $r \leq l$. In this case, $\mathbf{U}$ and $\mathbf{V}$ in (2.22) can be partitioned as $\mathbf{U} = [\mathbf{U}_1, \mathbf{U}_2]$ and $\mathbf{V} = [\mathbf{V}_1, \mathbf{V}_2]$, where $\mathbf{U}_1 = [\mathbf{u}_1, \ldots, \mathbf{u}_r]$ and $\mathbf{V}_1 = [\mathbf{v}_1, \ldots, \mathbf{v}_r]$ have r columns. Then

$$\mathbf{A} = [\mathbf{U}_1 \quad \mathbf{U}_2] \begin{bmatrix} \mathbf{\Sigma}_r & \mathbf{0} \\ \mathbf{0} & \mathbf{0} \end{bmatrix} [\mathbf{V}_1 \quad \mathbf{V}_2]^T = \mathbf{U}_1\mathbf{\Sigma}_r\mathbf{V}_1^T = \sum_{i=1}^{r} \sigma_i \mathbf{u}_i \mathbf{v}_i^T. \tag{2.24}$$

While equation (2.22) presents the full decomposition with square matrices **U** and **V**, equation (2.24) gives the *reduced* or the *economy* version of the SVD. In MATLAB, the economy size decomposition of matrix **A** is achieved using `[U, S, V] = svd(A, 0)`, where `S` denotes $\boldsymbol{\Sigma}_r$.

If dimension of **A** is $m \times n$ with $m > n$, then only the first n columns of **U** are computed and the size of $\boldsymbol{\Sigma}_r$ is $n \times n$. For $n \geq m$, `svd(A,0)` is equivalent to `svd(A)`. The function `svd(A, 'econ')` also produces the economy size decomposition. If **A** is $m \times n$ with $m \geq n$, then it is equivalent to `svd(A,0)`. For $m < n$, only the first m columns of **V** are computed and $\boldsymbol{\Sigma}_r$ is $m \times m$.

Example 10 Using `svd` function, find an SVD (full and economy decomposition) for the given matrix **A** in Example 9.

Solution This is done as:

```
>> A=[0 1;1 1;1 0]
A =
      0      1
      1      1
      1      0
>> [U,S,V] = svd(A)
U =
   -0.4082    0.7071    0.5774
   -0.8165    0.0000   -0.5774
   -0.4082   -0.7071    0.5774

S =
    1.7321         0
         0    1.0000
         0         0
V =
   -0.7071   -0.7071
   -0.7071    0.7071

>> [U,S,V] = svd(A,0)

U =
   -0.4082    0.7071
   -0.8165    0.0000
   -0.4082   -0.7071
S =
    1.7321         0
         0    1.0000
V =
   -0.7071   -0.7071
   -0.7071    0.7071
```

□

Example 11 Compute the reduced version of SVD for matrix $\mathbf{B} = \begin{bmatrix} 2 & 2 \\ 1 & 1 \end{bmatrix}$.

Solution **B** is of rank 1. First, we need to compute $\mathbf{B}^T\mathbf{B}$ and its eigenvectors.

$$\mathbf{B}^T\mathbf{B} = \begin{bmatrix} 5 & 5 \\ 5 & 5 \end{bmatrix}$$

The characteristic equation is $\det(\mathbf{B}^T\mathbf{B} - \lambda\mathbf{I}_2) = (5-\lambda)^2 - 25 = 0$. Then, the $\mathbf{B}^T\mathbf{B}$ has two eigenvalues $\lambda_1 = 10$ and $\lambda_2 = 0$. So, the corresponding singular values are $\sigma_1 = \sqrt{10}$ and $\sigma_2 = 0$. For reduced SVD, the nonzero singular value must be considered (i.e., σ_1). The corresponding eigenvector of λ_1 is obtained as follows:

$$(\mathbf{B}^T\mathbf{B} - \lambda_1\mathbf{I}_2)\mathbf{x}_1 = \begin{bmatrix} -5 & 5 \\ 5 & -5 \end{bmatrix}\mathbf{x}_1 = 0 \Rightarrow \mathbf{x}_1 = \begin{bmatrix} 1 \\ 1 \end{bmatrix}$$

Normalized form of $\mathbf{x_1}$ gives $\mathbf{v}_1 = \begin{bmatrix} \frac{1}{\sqrt{2}} \\ \frac{1}{\sqrt{2}} \end{bmatrix}$. Now, to find $\mathbf{u}_1$, we multiply $\mathbf{v}_1$ by **B** as:

$$\mathbf{B}\mathbf{v}_1 = \begin{bmatrix} 2 & 2 \\ 1 & 1 \end{bmatrix}\begin{bmatrix} \frac{1}{\sqrt{2}} \\ \frac{1}{\sqrt{2}} \end{bmatrix} = \begin{bmatrix} 2\sqrt{2} \\ \sqrt{2} \end{bmatrix}.$$

From (2.24), the unit vector $\mathbf{u}_1$ can be obtained as follows:

$$\mathbf{u}_1 = \frac{\mathbf{B}\mathbf{v}_1}{\sigma_1} = \frac{1}{\sqrt{10}}\begin{bmatrix} 2\sqrt{2} \\ \sqrt{2} \end{bmatrix} = \begin{bmatrix} \frac{2}{\sqrt{5}} \\ \frac{1}{\sqrt{5}} \end{bmatrix}.$$

This gives the reduced SVD as:

$$\mathbf{B} = \mathbf{u}_1\sigma_1\mathbf{v}_1^T = \begin{bmatrix} \frac{2}{\sqrt{5}} \\ \frac{1}{\sqrt{5}} \end{bmatrix}\sqrt{10}\begin{bmatrix} \frac{1}{\sqrt{2}} & \frac{1}{\sqrt{2}} \end{bmatrix}.$$

□

2.10 Quadratic Form

A *quadratic form* is a polynomial function in which the highest-degree term is of degree 2. A general quadratic function $f : \mathbb{R}^n \to \mathbb{R}$ is in the form of:

$$f(\mathbf{x}) = \mathbf{x}^T\mathbf{Q}\mathbf{x} = \sum_{i=1}^{n}\sum_{j=1}^{n} x_i q_{ij} x_j \tag{2.25}$$

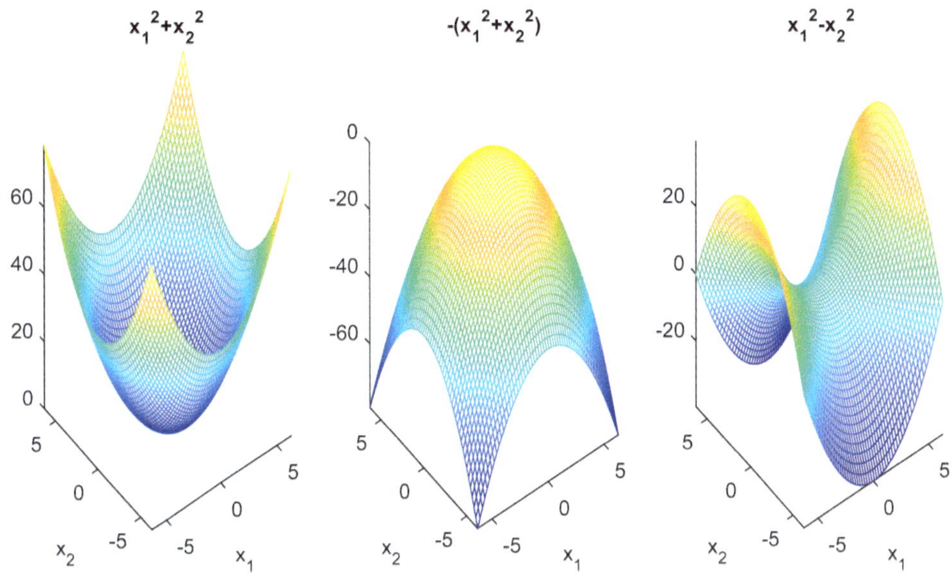

Fig. 2.1 Curvature of quadratic functions

where $\mathbf{Q}$ is a square matrix in $\mathbb{R}^{n\times n}$ with $\{q_{ij}\}$ entries and $\mathbf{x} \in \mathbb{R}^n$. Without loss of generality, $\mathbf{Q}$ is assumed to be symmetric. If not, it would be possible to replace it with an equivalent symmetric matrix $\mathbf{Q}_0 = \frac{1}{2}(\mathbf{Q}+\mathbf{Q}^T)$ such that $f(\mathbf{x}) = \mathbf{x}^T\mathbf{Q}\mathbf{x} = \mathbf{x}^T\mathbf{Q}_0\mathbf{x}$.

Example 12 Find the quadratic form for $\mathbf{Q} = \begin{bmatrix} 3 & 4 \\ 6 & 2 \end{bmatrix}$.

Solution Let $\mathbf{x} = \begin{bmatrix} x_1 \\ x_2 \end{bmatrix} \Rightarrow \mathbf{x}^T\mathbf{Q}\mathbf{x} = [x_1\ x_2]\begin{bmatrix} 3 & 4 \\ 6 & 2 \end{bmatrix}\begin{bmatrix} x_1 \\ x_2 \end{bmatrix} = 3x_1^2 + 10x_1x_2 + 2x_2^2$. Moreover, let $\mathbf{Q}_0 = \frac{1}{2}(\mathbf{Q}+\mathbf{Q}^T) = \begin{bmatrix} 3 & 5 \\ 5 & 2 \end{bmatrix} \Rightarrow \mathbf{x}^T\mathbf{Q}_0\mathbf{x} = 3x_1^2 + 10x_1x_2 + 2x_2^2$ that is the same as $\mathbf{x}^T\mathbf{Q}\mathbf{x}$. □

Following definitions are used to classify quadratic form functions. A quadratic form $\mathbf{x}^T\mathbf{Q}\mathbf{x}$ in which $\mathbf{Q} = \mathbf{Q}^T$ is said to be:

- Positive definite, denoted by $\mathbf{Q} > 0$, if $\mathbf{x}^T\mathbf{Q}\mathbf{x} > 0$ for all $\mathbf{x} \neq 0$.
- Positive semi-definite, denoted by $\mathbf{Q} \geq 0$, if $\mathbf{x}^T\mathbf{Q}\mathbf{x} \geq 0$ for all $\mathbf{x} \neq 0$.
- Negative definite, denoted by $\mathbf{Q} < 0$, if $\mathbf{x}^T\mathbf{Q}\mathbf{x} < 0$ for all $\mathbf{x} \neq 0$.
- Negative semi-definite, denoted by $\mathbf{Q} \leq 0$, if $\mathbf{x}^T\mathbf{Q}\mathbf{x} \leq 0$ for all $\mathbf{x} \neq 0$.
- Otherwise, $\mathbf{x}^T\mathbf{Q}\mathbf{x}$ is indefinite.

To illustrate this classification, the following code in MATLAB generates the graphs shown in the Fig. 2.1 using three typical functions:

```
f1 = @(x1, x2) x1^2 + x2^2;
f2 = @(x1, x2) -(x1^2 + x2^2);
```

```
f3 = @(x1, x2) x1^2 - x2^2;
subplot(1,3,1), ezmesh(f1)
subplot(1,3,2), ezmesh(f2)
subplot(1,3,3), ezmesh(f3)
```

1. $f(\mathbf{x}) = x_1^2 + x_2^2 = \mathbf{x}^T\mathbf{Q}\mathbf{x}$, , where $\mathbf{Q} = \begin{bmatrix} 1 & 0 \\ 0 & 1 \end{bmatrix}$. For every $\mathbf{x} \neq 0$, $f(\mathbf{x}) > 0$. Therefore, $f(\mathbf{x})$ is a positive definite function. As seen, the curvature is upward.
2. $f(\mathbf{x}) = -(x_1^2 + x_2^2) = \mathbf{x}^T\mathbf{Q}\mathbf{x}$ where $\mathbf{Q} = \begin{bmatrix} -1 & 0 \\ 0 & -1 \end{bmatrix}$. For every $\mathbf{x} \neq 0$, $f(\mathbf{x}) < 0$. Therefore, $f(\mathbf{x})$ is a negative definite function. The curvature is downward.
3. $f(\mathbf{x}) = x_1^2 - x_2^2 = \mathbf{x}^T\mathbf{Q}\mathbf{x}$, where $\mathbf{Q} = \begin{bmatrix} 1 & 0 \\ 0 & -1 \end{bmatrix}$. This function is positive along x_1 and negative along x_2. Therefore, it is indefinite.

Consider the statement in Theorem 14 to determine whether a matrix $\mathbf{Q}$ is positive or negative definite.

Theorem 14 *A symmetric matrix* $\mathbf{Q}$ *is positive definite (or positive semi-definite) if and only if all eigenvalues of* $\mathbf{Q}$ *are positive (non-negative).*

Proof Recall that for a symmetric matrix $\mathbf{Q}$, the diagonalization is in the form of $\mathbf{Q} = \mathbf{V}\mathbf{D}\mathbf{V}^T$ as in (2.21), where $\mathbf{D}$ contains the eigenvalues on the main diagonal. Now, for any $\mathbf{x} \in \mathbb{R}^n$ and symmetric $\mathbf{Q} = \mathbf{V}\mathbf{D}\mathbf{V}^T$, consider

$$\mathbf{x}^T\mathbf{Q}\mathbf{x} = \mathbf{x}^T\mathbf{V}\mathbf{D}\mathbf{V}^T\mathbf{x} = (\mathbf{V}^T\mathbf{x})^T\mathbf{D}(\mathbf{V}^T\mathbf{x})$$

Let $\mathbf{y} \triangleq \mathbf{V}^T\mathbf{x} \Rightarrow \mathbf{x}^T\mathbf{Q}\mathbf{x} = \mathbf{y}^T\mathbf{D}\mathbf{y} = \sum_{i=1}^n \lambda_i y_i^2$. Considering this derivation, if $\mathbf{x}^T\mathbf{Q}\mathbf{x} \geq 0$ for all $\mathbf{x} \neq 0$, we drive $\sum_{i=1}^n \lambda_i y_i^2 \geq 0$ for all y_i's, which in turn implies $\lambda_i \geq 0$ for all i.

Conversely, $\lambda_i \geq 0$ for all i implies $\sum_{i=1}^n \lambda_i y_i^2 \geq 0$ for all y_i's and consequently $\mathbf{x}^T\mathbf{Q}\mathbf{x} \geq 0$ for all $\mathbf{x} \neq 0$. □

Notice that Theorem 14 can be extended to the case of negative definite matrices. If $\mathbf{Q}$ is positive definite, then $-\mathbf{Q}$ is negative definite. Therefore, a symmetric matrix $\mathbf{Q}$ is negative definite (or negative-semi definite) if and only if all eigenvalues of $\mathbf{Q}$ are negative (nonpositive).

2.11 Gradient

Gradient is the extension of the derivative operation from a single variable to a multivariable vector. The gradient of a function of several variables is a vector-valued function whose components are the partial derivatives of the function. More precisely, the gradient of a scalar and differentiable function $f : \mathbb{R}^n \rightarrow \mathbb{R}$ is denoted by $\nabla f(\mathbf{x})$ and is defined as:

$$\nabla f(\mathbf{x}) = \begin{bmatrix} \frac{\partial f(\mathbf{x})}{\partial x_1} \\ \frac{\partial f(\mathbf{x})}{\partial x_2} \\ \vdots \\ \frac{\partial f(\mathbf{x})}{\partial x_n} \end{bmatrix} \tag{2.26}$$

where $\frac{\partial f(\mathbf{x})}{\partial x_i}$ is the partial derivative of f with respect to x_i and indicates the rate of increase of f in the direction of x_i.

Example 13 Find the gradient of $f(\mathbf{x}) = x_1^2 + x_3^2 + 2x_1x_2 + 2x_1x_3 + 2x_2x_3$.

Solution Using partial derivatives:

$$\nabla f(\mathbf{x}) = \begin{bmatrix} 2x_1 + 2x_2 + 2x_3 \\ 2x_1 + 2x_3 \\ 2x_1 + 2x_2 + 2x_3 \end{bmatrix}.$$

Moreover, using the MATLAB commands, we derive

```
>> syms x1 x2 x3
>> f = x1^2 + x3^2 + 2*x1*x2 + 2*x1*x3 + 2*x2*x3;
>> gradf = jacobian(f,[x1,x2, x3])

gradf =

[ 2*x1 + 2*x2 + 2*x3, 2*x1 + 2*x3, 2*x1 + 2*x2 + 2*x3]
```

□

Some well-known forms of gradient functions are:

- $f(\mathbf{x}) = \mathbf{b}^T\mathbf{x}$, $\mathbf{b} \in \mathbb{R}^n \Rightarrow \nabla f(\mathbf{x}) = \mathbf{b}$
- $f(\mathbf{x}) = \mathbf{x}^T\mathbf{Q}\mathbf{x}$, $\mathbf{Q} \in \mathbb{R}^{n\times n}$ is symmetric $\Rightarrow \nabla f(\mathbf{x}) = 2\mathbf{Q}\mathbf{x}$.

2.12 Hessian Matrix

The Hessian of a scalar-valued function is a square matrix containing the second-order partial derivative functions. This matrix denotes the local curvature of the function. Given $f : \mathbb{R}^n \to \mathbb{R}$, if $\nabla f(\mathbf{x})$ is differentiable, we say that $f(\mathbf{x})$ is twice differentiable and write the Hessian matrix as:

$$\mathbf{H} = \nabla^2 f(\mathbf{x}) = \begin{bmatrix} \frac{\partial^2 f}{\partial x_1^2} & \frac{\partial^2 f}{\partial x_2 \partial x_1} & \cdots & \frac{\partial^2 f}{\partial x_n \partial x_1} \\ \frac{\partial^2 f}{\partial x_1 \partial x_2} & \frac{\partial^2 f}{\partial x_2^2} & \cdots & \frac{\partial^2 f}{\partial x_n \partial x_2} \\ \vdots & \vdots & \cdots & \vdots \\ \frac{\partial^2 f}{\partial x_1 \partial x_n} & \frac{\partial^2 f}{\partial x_2 \partial x_n} & \cdots & \frac{\partial^2 f}{\partial x_n^2} \end{bmatrix}. \tag{2.27}$$

Due to the commutative property of partial derivatives, the Hessian matrix is symmetric. Furthermore:

- $f(\mathbf{x}) = \mathbf{b}^T\mathbf{x}, \mathbf{b} \in \mathbb{R}^n \Rightarrow \mathbf{H} = \mathbf{0}$
- $f(\mathbf{x}) = \mathbf{x}^T\mathbf{Q}\mathbf{x}$, $\mathbf{Q}$ is symmetric $\Rightarrow \mathbf{H} = 2\mathbf{Q}$.

Example 14 Find the Hessian of $f(\mathbf{x}) = x_1^2 + x_3^2 + 2x_1x_2 + 2x_1x_3 + 2x_2x_3$.

Solution Using the result of Example 13 and taking the second-order partial derivatives:

$$\mathbf{H} = \begin{bmatrix} 2 & 2 & 2 \\ 2 & 0 & 2 \\ 2 & 2 & 2 \end{bmatrix}$$

Alternatively, we can write the quadratic function as $f(\mathbf{x}) = \mathbf{x}^T\mathbf{Q}\mathbf{x}$, where $\mathbf{Q} = \begin{bmatrix} 1 & 1 & 1 \\ 1 & 0 & 1 \\ 1 & 1 & 1 \end{bmatrix}$. Therefore, $\nabla f(\mathbf{x}) = 2\mathbf{Q}\mathbf{x}$ and $\mathbf{H} = 2\mathbf{Q}$, which is the same as above derivation. Finally, the MATLAB commands are as following:

```
>> syms x1 x2 x3
>> f = x1^2 + x3^2 + 2*x1*x2 + 2*x1*x3 + 2*x2*x3;
>> hessian(f, [x1,x2, x3])

ans =

[ 2, 2, 2]
[ 2, 0, 2]
[ 2, 2, 2]
```

□

2.13 Supremum and Infimum

Supremum and infimum are two operations that are used frequently in the literature of optimization. Before introducing these operations, consider definitions of upper bound and lower bound terms in Definition 10.

Definition 10 The real number M is called an upper bound of the real set $\mathcal{S} \subset \mathbb{R}$ if $x \leq M$ for all $x \in \mathcal{S}$. Similarly, the real number m is called a lower bound of $\mathcal{S} \subset \mathbb{R}$ if $x \geq m$ for all $x \in \mathcal{S}$.

Following this definition, the *supremum* of $\mathcal{S}$ is its least upper bound, denoted by sup $\mathcal{S}$. The supremum is also called the *maximum* of $\mathcal{S}$ if it belongs to $\mathcal{S}$, denoted by max $\mathcal{S}$. Similarly, the infimum of $\mathcal{S}$ is its greatest lower bound, denoted by inf $\mathcal{S}$. The infimum is also called the minimum of $\mathcal{S}$ if it belongs to $\mathcal{S}$, denoted by min $\mathcal{S}$.

Example 15 Consider the set $\mathcal{S} = \{\frac{1}{n} | n \in \mathbb{N}\}$, where $\mathbb{N}$ is the set of positive integers. Find sup, inf, max, and min of $\mathcal{S}$.

Solution Inspecting the elements in $\mathcal{S}$, it is seen that sup $\mathcal{S} = 1$ that belongs to $\mathcal{S}$, so max $\mathcal{S} = 1$. On the other hand, inf $\mathcal{S} = 0$ that does not belong to $\mathcal{S}$, so it is not a minimum. Indeed, there is no minimum for $\mathcal{S}$. □

Following the definitions of supremum and infimum for real sets, these operations are also defined for every function $f : \Omega \to \mathbb{R}$ as:

$$\sup_{\mathbf{x}\in\Omega} f(\mathbf{x}) = \sup \{f(\mathbf{x}) | \mathbf{x} \in \Omega\} \tag{2.28}$$

and

$$\inf_{\mathbf{x}\in\Omega} f(\mathbf{x}) = \inf \{f(\mathbf{x}) | \mathbf{x} \in \Omega\} \tag{2.29}$$

where Ω is the domain of f.

2.14 Level Set

For a real-valued function $f : \mathbb{R}^n \to \mathbb{R}$, the level set at a level $c \in \mathbb{R}$ is the set of points or vectors satisfying:

$$\mathcal{S}_c = \{\mathbf{x} \in \mathbb{R}^n | f(\mathbf{x}) = c\}. \tag{2.30}$$

In general, a level set is the set of all real-valued roots of an equation with n variables $x_1, x_2, \ldots, x_n$. If $n = 2$, the level set is called *curve*, and if $n = 3$, the level set is called *surface*, and for higher values of n, it is called *hyper-surface*.

As an example, the following code in MATLAB plots the level sets of the graph of $f(\mathbf{x}) = x_1^2 + x_2^2$ for a number of constant values c in Fig. 2.2 ($f(\mathbf{x}) = c$). As shown, level sets appear in circular forms for this function. Each curve is labeled with a value indicating its level c.

```
delta = 0.25;
[x1, x2] = meshgrid(-3:delta:3);
f = x1.^2 + x2.^2;
c = contour(x1, x2, f); clabel(c);
xlabel('x_1'), ylabel('x_2')
```

In addition to level sets, sublevel sets $\mathbf{S}_c^-$ and superlevel sets $\mathcal{S}_c^+$ are also defined as:

$$\mathcal{S}_c^- = \{\mathbf{x} \in \mathbb{R}^n | f(\mathbf{x}) \leq c\} \tag{2.31}$$

and

$$\mathcal{S}_c^+ = \{\mathbf{x} \in \mathbb{R}^n | f(\mathbf{x}) \geq c\}. \tag{2.32}$$

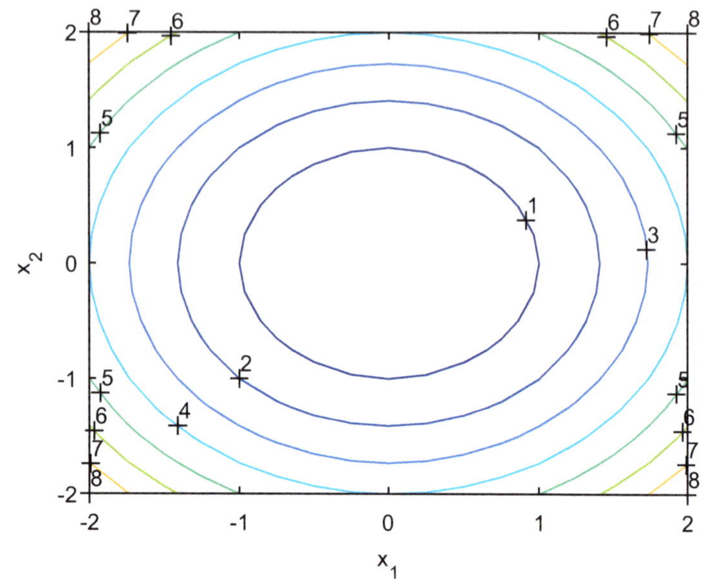

Fig. 2.2 Level sets of $f(\mathbf{x}) = x_1^2 + x_2^2$

Sublevel and superlevel sets are important in optimization theory as the feasible region of inequality constraints appeared in these forms and they determine the boundary of the optimization feasible region. Following, in Theorem 15, the relation between level sets and gradient vectors is discussed.

Theorem 15 *Consider level set* $S = \{\mathbf{x}|f(\mathbf{x}) = c\}$ *and* $\mathbf{x}_0 \in S$. *Gradient vector* $\nabla f(\mathbf{x}_0)$ *is orthogonal to* S *at* $\mathbf{x}_0$.

Proof Let S be a surface represented by $f(\mathbf{x}) = c$ as in Fig. 2.3 and f is differentiable. Every curve C on S passing through a point $\mathbf{x}_0$ can be shown as a parametric representation with parameter t as $\mathbf{x} = \mathbf{g}(t)$, where $\mathbf{g} : \mathbb{R} \to \mathbb{R}^n$ and consequently $f(\mathbf{g}(t)) = c$. The tangent vector of every curve C on S is obtained as:

$$D\mathbf{g}(t) = \frac{d\mathbf{g}(t)}{dt} = \begin{bmatrix} \frac{\partial x_1}{\partial t} \\ \frac{\partial x_2}{\partial t} \\ \vdots \\ \frac{\partial x_n}{\partial t} \end{bmatrix} \triangleq \mathbf{v}. \tag{2.33}$$

All tangent vectors $\mathbf{v}$ passing through $\mathbf{x}_0$ form a tangent plane as in Fig. 2.3. Recall that the normal vector of this plane is perpendicular to all tangent vectors. Taking derivative of $f(\mathbf{g}(t)) = c$ with respect to t and using the chain rule, we derive

$$\frac{df}{dt} = \frac{\partial f}{\partial x_1}.\frac{\partial x_1}{\partial t} + \cdots + \frac{\partial f}{\partial x_n}.\frac{\partial x_n}{\partial t} = \nabla f(\mathbf{x})^T.\mathbf{v} = 0 \Rightarrow \nabla f(\mathbf{x}) \perp \mathbf{v}. \tag{2.34}$$

Therefore, it is concluded that $\nabla f(x_0)$ is orthogonal to S at $\mathbf{x}_0$. $\square$

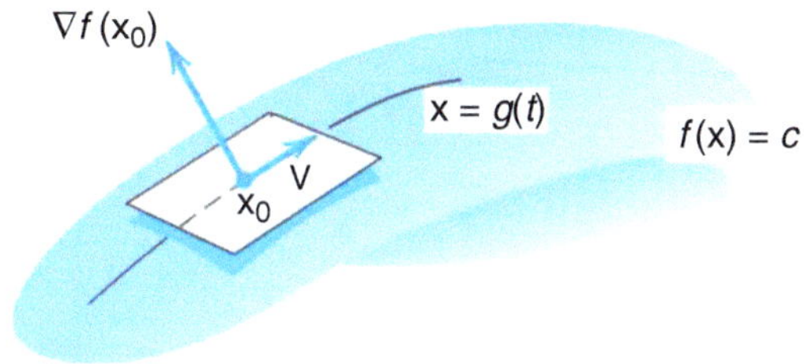

Fig. 2.3 $\nabla f(\mathbf{x}_0)$ is orthogonal to tangent vector $\mathbf{v}$ [3]

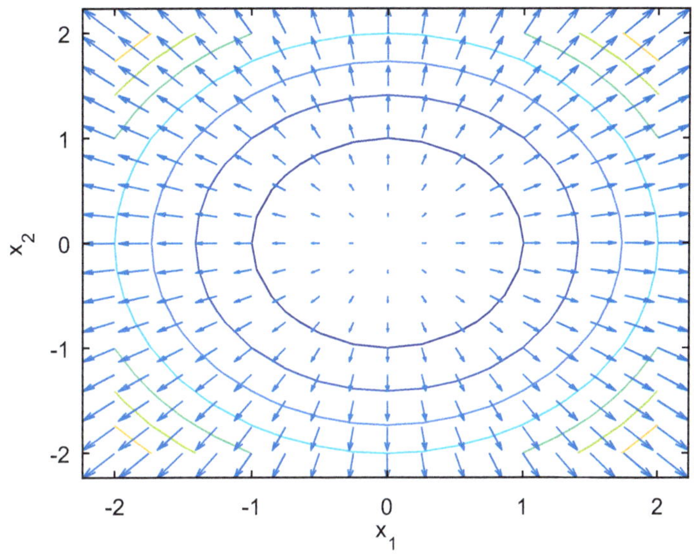

Fig. 2.4 Gradient vectors of $f(\mathbf{x}) = x_1^2 + x_2^2$

As an example, the gradient vectors of $f(\mathbf{x}) = x_1^2 + x_2^2$ are plotted on the level sets of its graph in Fig. 2.4. As shown, gradient vector at each point is perpendicular to the corresponding level set. Moreover, the length of each vector corresponds to its magnitude. This is done in MATLAB as:

```
delta = 0.25;
[x1, x2] = meshgrid(-3:delta:3);
f = x1.^2 + x2.^2;
[fx1, fx2] = gradient(f, delta, delta);
 contour(x1, x2, f); xlabel('x_1'), ylabel('x_2')
 hold on, quiver(x1, x2, fx1, fx2, 1)
```

2.15 Directional Derivative

Recall that in the definition of gradient in Sect. 2.11, $\frac{\partial f}{\partial x_i}$ was the rate of increase of f along x_i. Directional derivative is an extension of this statement in an arbitrary direction. In particular, directional derivative of $f : \mathbb{R}^n \to \mathbb{R}$ in the direction of $\mathbf{d} = \begin{bmatrix} d_1 \\ \vdots \\ d_n \end{bmatrix}$ is defined by:

$$\frac{\partial f}{\partial \mathbf{d}} = \lim_{\alpha \to 0} \frac{f(\mathbf{x} + \alpha \mathbf{d}) - f(\mathbf{x})}{\alpha}. \tag{2.35}$$

Using the chain rule:

$$\frac{\partial f}{\partial \mathbf{d}} = d_1 \frac{\partial f}{\partial x_1} + \cdots + d_n \frac{\partial f}{\partial x_n} = \nabla f(\mathbf{x})^T$$

$$\mathbf{d} = \langle \nabla f(\mathbf{x}), \mathbf{d} \rangle.$$

Note that if $\|\mathbf{d}\| = 1$, then $\frac{\partial f}{\partial \mathbf{d}}$ is the rate of increase of f at $\mathbf{x}$ in the direction of $\mathbf{d}$.

Example 16 Compute the directional derivative of $f(\mathbf{x}) = x_1 x_2 x_3$ in the direction of $\mathbf{d} = [2/3,\ 1/3,\ 2/3]^T$.

Solution $\nabla f(\mathbf{x})^T \mathbf{d} = [x_2 x_3, x_1 x_3, x_1 x_2] \begin{bmatrix} 2/3 \\ 1/3 \\ 2/3 \end{bmatrix} = \frac{2x_2 x_3}{3} + \frac{x_1 x_3}{3} + \frac{2x_1 x_2}{3}$. □

The concept of directional derivative is important due to its application in optimization theory pointed out by Theorem 16.

Theorem 16 *The maximum increase rate of $f : \mathbb{R}^n \to \mathbb{R}$ at point $\mathbf{x}_0$ occurs in the direction of the gradient vector, i.e., $\nabla f(\mathbf{x}_0)$.*

Proof The directional derivative of f at point $\mathbf{x}_0$ in the direction of $\mathbf{d}$ is $\nabla f(\mathbf{x}_0)^T \mathbf{d}$ that is equivalent to $\|\nabla f(\mathbf{x}_0)\| \|\mathbf{d}\| \cos(\theta)$, where θ is the angle between $\nabla f(\mathbf{x}_0)$ and $\mathbf{d}$ vectors. Consequently, $\nabla f(\mathbf{x}_0)^T \mathbf{d}$ would be maximized if $\theta = 0$ or equivalently $\mathbf{d} = \alpha \nabla f(\mathbf{x})$ for some $\alpha > 0$. Similarly, $\nabla f(\mathbf{x}_0)^T \mathbf{d}$ would be minimized if $\theta = \pi$ or equivalently $\mathbf{d} = -\alpha \nabla f(\mathbf{x})$ for some $\alpha > 0$. □

The derivation in Theorem 16 is of high importance in optimization theory as we always need to know the value and the direction of maximum or minimum rate of increase at a given point in order to reach the optimal solution. This statement is employed frequently in the rest of the book.

2.16 Summary

Linear algebra as a main prerequisite of optimization in engineering has been reviewed in this chapter. Beginning with basic terms of vector and matrix spaces, some well-known characteristics of these spaces including rank, determinant, minor, inverse, inner product, and norm have been reviewed. Using the properties of these spaces, the number of solutions of a set of linear equations has been addressed. Furthermore, eigenvalues, eigenvectors, singular values, and diagonalization as more useful topics in the matrix space have been also discussed. The rest of the chapter addressed quadratic form functions and a number of characteristics for general form functions such as gradient, Hessian, level set, and directional derivative. Throughout the chapter, numerical examples have been done, and more importantly the solutions have been provided with MATLAB scripts.

2.17 Problems

Problem 1 A set of vectors $\{\mathbf{a}_1, \ldots, \mathbf{a}_N\} \in \mathbb{R}^n$ is said to be orthonormal if:

$$\mathbf{a}_i^T \mathbf{a}_j = \begin{cases} 1 & if \quad i = j \\ 0 & if \quad i \neq j \end{cases} \quad \text{for all } i,\ j.$$

(a) Show that any orthonormal set is linearly independent.
(b) Let $\mathbf{b} \in \mathbb{R}^n$ be represented as $\mathbf{b} = \sum_{i=1}^{N} \alpha_i \mathbf{a}_i$. Derive coefficients α_i's.

Problem 2 Find the rank of the following matrices using column operations:

(a) $\mathbf{A} = \begin{bmatrix} -1 & 2 & 0 \\ 5 & 1 & -2 \\ 6 & 3 & 4 \end{bmatrix}$

(b) $\mathbf{A} = \begin{bmatrix} 3 & 2 & 2 & -5 \\ 0.6 & 1.5 & 1.5 & -5.4 \\ 1.2 & -0.3 & -0.3 & 2.4 \end{bmatrix}$

Problem 3 Find the number of solutions for linear equations with the following augmented matrices $\tilde{\mathbf{A}} = [\mathbf{A}\ \mathbf{b}]$:

(a) $\tilde{\mathbf{A}} = \begin{bmatrix} 2 & -1 & 3 & 0 \\ -1 & 4 & -1 & 10 \\ 0 & 10 & 15 & 50 \\ 20 & 10 & 1 & 40 \end{bmatrix}$

(b) $\tilde{\mathbf{A}} = \begin{bmatrix} 4 & 1 & 0 & 4 \\ 5 & -3 & 1 & 2 \\ -9 & 2 & -1 & 5 \end{bmatrix}$

Problem 4 Find eigenvalues and eigenvectors of:

(a) $\mathbf{A} = \begin{bmatrix} -5 & 0 \\ 0 & 3 \end{bmatrix}$

(b) $\mathbf{A} = \begin{bmatrix} -1 & 2 \\ 5 & 2 \end{bmatrix}$

Problem 5 Diagonalize the following matrices in the form of $\mathbf{A} = \mathbf{V}\mathbf{D}\mathbf{V}^{-1}$ or $\mathbf{A} = \mathbf{V}\mathbf{D}\mathbf{V}^{T}$:

(a) $\mathbf{A} = \begin{bmatrix} 1 & -3 \\ 4 & 10 \end{bmatrix}$

(b) $\mathbf{A} = \begin{bmatrix} 4 & 3 & 6 \\ 3 & -8 & 15 \\ 6 & 15 & 10 \end{bmatrix}$

Problem 6 Find SVD for the following matrices:

(a) $\mathbf{A} = \begin{bmatrix} 1 & 2 & 1 \\ 2 & 3 & 2 \\ 1 & 2 & 1 \end{bmatrix}$.

(b) $\mathbf{A} = \begin{bmatrix} 0 & 1 & 0 & 0 \\ 0 & 0 & 4 & 0 \\ 0 & 0 & 0 & 9 \\ 0 & 0 & 0 & 0 \end{bmatrix}$.

Problem 7 Let $\mathbf{B}$ be an $m \times n$ matrix in which $m \geq n$. Show that the singular values of $\mathbf{A} = \begin{bmatrix} \mathbf{I}_n \\ \mathbf{B} \end{bmatrix}$ can be calculated from $\sqrt{1 - \sigma_i^2}$, $1 \leq i \leq n$.

Problem 8 Check out the definiteness of the following quadratic expressions:

(a) $f(x_1, x_2) = x_1^2 + 6x_1x_2 + x_2^2$.
(b) $f(x_1, x_2) = 2x_1^2 + 6x_1x_2 - 2x_2^2$.
(c) $f(x_1, x_2, x_3) = x_1^2$.
(d) $f(x_1, x_2, x_3) = x_1^2 + x_3^2 + 2x_1x_2 + 2x_1x_3 + 2x_2x_3$.

Problem 9 Using MATLAB, plot the graph, level sets and gradient vectors of:

$$f(x_1, x_2) = \frac{x_1^4 + 2x_1^3x_2 - 6x_1^2x_2^2 + x_2^4}{x_1^4 + x_2^4 + 1}$$

for $-3 \leq x_1, x_2 \leq 3$.

Problem 10 Show that row rank and column rank are equal for the following matrix:

$$\mathbf{A} = \begin{bmatrix} 1 & 0 & 2 & 2 \\ -2 & 42 & 24 & 4 \\ 7 & -21 & 0 & -5 \end{bmatrix}$$

Problem 11

(a) Any real symmetric matrix $\mathbf{A} \in \mathbf{R}^{n \times n}$ has a set of eigenvectors $[v_1, \ldots, v_n]$ that are mutually orthogonal, show that $\mathbf{A} = \sum_{i=1}^{n} \lambda_i v_i v_i^T$, where $\{\lambda_i\}_{i=1}^{n}$ are eigenvalues.
(b) Using the result in part (a), show that symmetric matrix $\mathbf{A}$ is positive definite if all eigenvalues are positive.

Problem 12 Let $\lambda_{\min}$ and $\lambda_{\max}$ be the minimum and maximum eigenvalues of a symmetric matrix $\mathbf{A}$, respectively, and normalized eigenvectors.

(a) Show that:

$$\lambda_{\min}\|\mathbf{x}\|^2 \leq \mathbf{x}^T\mathbf{A}\mathbf{x} \leq \lambda_{\max}\|\mathbf{x}\|^2.$$

for any $\mathbf{x} \in \mathbb{R}^n$.
(b) If $\mathbf{A}$ is symmetric and $\lambda_{\max} \leq \gamma$, then show that $\mathbf{A} - \gamma\mathbf{I}$ is negative semi-definite.

Problem 13 Prove that the columns of $\mathbf{A} = [\mathbf{a}_1, \ldots, \mathbf{a}_n]$ are linearly dependent if and only if $\det \mathbf{A} = 0$. In other words, the columns of $\mathbf{A}$ are linearly independent if and only if $\det \mathbf{A} \neq 0$.

Problem 14 Prove

(a) Sum of all entries of an $n \times n$ positive definite matrix is positive.
(b) All entries in the main diagonal of an $n \times n$ positive definite matrix is positive.

References

1. E.K.P. Chong, S.H. Zak, *An Introduction to Optimization* (Wiley, Hoboken, 2013)
2. G. Strang, *Linear Algebra and Its Applications* (Cengage Learning, Stamford, 2014)
3. E. Kreyszig, *Advanced Engineering Mathematics* (Wiley, Hoboken, 2011)
4. C. D. Meyer, *Matrix Analysis and Applied Linear Algebra* (Siam, USA, 2000)

Set Constrained Optimization

3

Abstract

A basic form of an optimization problem is to minimize or maximize an objective function of real variables in a given set. This kind of optimization is usually known as set constrained optimization. This chapter discusses some optimization concepts and conceptual optimality conditions of set constrained optimization. A number of basic and general optimization algorithms using iterative search methods such as steepest descent and Newton's methods are then introduced with extensive examples.

Keywords

Iterative search method · Least square optimization · Newton · Steepest descent · Unconstrained optimization

3.1 Set Constrained Optimization Problem

Consider the problem of minimizing a function over a given set. Mathematically, this statement can be expressed as an optimization problem in the form of:

$$\min_{\mathbf{x}} \ f(\mathbf{x}) \tag{3.1a}$$

$$\text{subject to } \ \mathbf{x} \in \Omega \tag{3.1b}$$

where $f : \mathbb{R}^n \to \mathbb{R}$ is the *objective function*, $\mathbf{x} \in \mathbb{R}^n$ is the *optimization variable*, and $\Omega \subset \mathbb{R}^n$ is the *feasible set* or constraint set. Any $\mathbf{x} \in \Omega$ is called a feasible solution. The optimal solution or optimal point of this problem is the value of $\mathbf{x} = \mathbf{x}^*$ which has the smallest value of f among all other vectors in the feasible set. In other words:

$$\mathbf{x}^* = \text{argmin}\{f(\mathbf{x})|\mathbf{x} \in \Omega\} \tag{3.2}$$

or equivalently:

$$f(\mathbf{x}^*) = \min\{f(\mathbf{x})|\mathbf{x} \in \Omega\}. \tag{3.3}$$

M. Fathi, H. Bevrani, *Optimization in Electrical Engineering*,
https://doi.org/10.1007/978-3-030-05309-3_3

Note that the problem of maximizing f is equivalent to minimizing $-f$. Therefore, in this book, optimization problems are mostly considered in the minimization form.

In problem (3.1), if $\Omega = \mathbb{R}^n$, then there is no restriction on the value of optimization variables, except that $\mathbf{x} \in \mathbb{R}^n$. Therefore, it is called an *unconstrained optimization* problem. These kind of problems arise directly in some applications but they also arise indirectly from reformulation of constrained optimization problems as unconstrained problems. Often, it is common to replace constraints of an optimization problem with penalized terms in the objective function and to solve it as an unconstrained problem.

This chapter investigates necessary and sufficient conditions for the optimal solution of set constrained optimization. Afterwards, a number of well-known algorithms are introduced to solve unconstrained optimization problems.

3.2 Local and Global Optimal Points

Prior to discussing optimality conditions, the following definitions on local and global optimal points are introduced.

Definition 1 (Local Minimizer) A point $\mathbf{x}^* \in \Omega$ is a local minimizer of f over Ω if there exists $\epsilon > 0$ such that $f(\mathbf{x}) \geq f(\mathbf{x}^*)$ for all $\mathbf{x} \in \Omega$ and $\|\mathbf{x} - \mathbf{x}^*\| \leq \epsilon$.

Definition 2 (Global Minimizer) A point $\mathbf{x}^* \in \Omega$ is the global minimizer of f over Ω if $f(\mathbf{x}) \geq f(\mathbf{x}^*)$ for all $\mathbf{x} \in \Omega$.

Definition 3 (Local Maximizer) A point $\mathbf{x}^* \in \Omega$ is a local maximizer of f over Ω if there exists $\epsilon > 0$ such that $f(\mathbf{x}) \leq f(\mathbf{x}^*)$ for all $\mathbf{x} \in \Omega$ and $\|\mathbf{x} - \mathbf{x}^*\| \leq \epsilon$.

Definition 4 (Global Maximizer) A point $\mathbf{x}^* \in \Omega$ is the global maximizer of f over Ω if $f(\mathbf{x}) \leq f(\mathbf{x}^*)$ for all $\mathbf{x} \in \Omega$.

In the first two definitions, if we replace $\geq$ with $>$, then we have strict local minimizer and strict global minimizer, respectively. This is also true for strict local maximizer and strict global maximizer if we replace $\leq$ with $<$ in last two definitions.

Note that global optimal points are also locally optimal. However, the converse is not true. Moreover, there could be several global optimal points, all with the same optimal value.

To clarify the mentioned definitions, consider the curve of a one-dimensional function plotted in Fig. 3.1. As shown, the points $x = 1$ and $x = 4$ are both local minimizers. Since $f(1) < f(4)$, the point $x = 1$ is the global minimizer. Moreover, the points $x = -3$, $x = 3$, and $x = 5$ are local

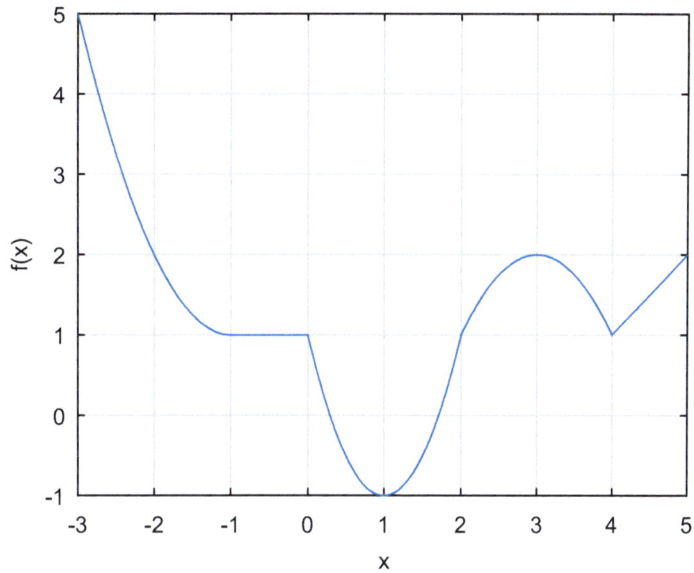

Fig. 3.1 Illustration of local and global optimal points

maximizers. Among them, $x = -3$ has the largest function value. Therefore, $x = -3$ is the global maximizer. Furthermore, all points in the range of $-1 \leq x \leq 0$ satisfy the definitions of both local minimum and maximum. However, they are not globally optimal.

3.3 Optimality Conditions

Recall from Chap. 2 that directional derivative $\frac{\partial f(\mathbf{x})}{\partial \mathbf{d}} = \mathbf{d}^T \nabla f(\mathbf{x})$ is the rate of increase of f at $\mathbf{x}$ in direction $\mathbf{d}$. Based on this statement, feasible direction is defined in Definition 5.

Definition 5 (Feasible Direction) At any point $\mathbf{x} \in \Omega$, a vector $\mathbf{d} \in \mathbb{R}^n$ is a feasible direction if $\mathbf{x} + \alpha\mathbf{d} \in \Omega$ for small enough $\alpha > 0$.

Feasible direction is used in the derivation of necessary and sufficient conditions of optimality conditions. These conditions are important in the design of algorithms for set constrained optimization [1, 2], which are stated in the upcoming theorems.

Theorem 1 (First-Order Necessary Condition) *Suppose that $f : \Omega \rightarrow \mathbb{R}$ is a differentiable function over Ω.*

(a) If $\mathbf{x}^*$ is a local minimizer of f over Ω, then $\mathbf{d}^T \nabla f(\mathbf{x}^*) \geq 0$ for all feasible directions $\mathbf{d}$ at $\mathbf{x}^*$.

(continued)

(b) If $\mathbf{x}^*$ is a local maximizer of f over Ω, then $\mathbf{d}^T\nabla f(\mathbf{x}^*) \leq 0$ for all feasible directions $\mathbf{d}$ at $\mathbf{x}^*$.

Proof The proof is given for part (a). Part (b) can be similarly proved by considering function $-f$. Let $\mathbf{x}^*$ be a local minimizer for any feasible direction $\mathbf{d}$, then:

$$f(\mathbf{x}^*) \leq f(\mathbf{x}^* + \alpha\mathbf{d}) \Rightarrow \frac{f(\mathbf{x}^* + \alpha\mathbf{d}) - f(\mathbf{x}^*)}{\alpha}$$
$$\geq 0 \Rightarrow \frac{\partial f(\mathbf{x}^*)}{\partial \mathbf{d}} \geq 0$$

or $\mathbf{d}^T\nabla f(\mathbf{x}^*) \geq 0$.

□

As a special case, note that if $\mathbf{x}^*$ is an interior point, the feasible direction is the whole of $\mathbb{R}^n$. So, for all $\mathbf{d}$, $\mathbf{d}^T\nabla f(\mathbf{x}^*) \geq 0$ and $-\mathbf{d}^T\nabla f(\mathbf{x}^*) \geq 0$, which implies

$$\nabla f(\mathbf{x}^*) = 0.$$

This means that the gradient vanishes at all local optimal points in the interior of the feasible region. However, the converse is not true. There could be points which are not locally optimal but the gradient is zero.

Example 1 For the problem:

$$\begin{aligned} &\text{minimize } 2x_1^2 + 0.5x_1x_2 + x_2^2 \\ &\text{subject to} \qquad x_1, x_2 \geq 0 \end{aligned}$$

verify if the first-order necessary condition is satisfied in the points $\mathbf{x} = \begin{bmatrix} 0.5 \\ 2 \end{bmatrix}$, $\mathbf{x} = \begin{bmatrix} 1.6 \\ 0 \end{bmatrix}$, $\mathbf{x} = \begin{bmatrix} 0 \\ 3.2 \end{bmatrix}$, $\mathbf{x} = \begin{bmatrix} 0 \\ 0 \end{bmatrix}$.

Solution The gradient is $\nabla f(\mathbf{x}) = \begin{bmatrix} 4x_1 + 0.5x_2 \\ 0.5x_1 + 2x_2 \end{bmatrix}$. Let's check out the sign of $\mathbf{d}^T\nabla f(\mathbf{x})$ for any feasible $\mathbf{d}$ at given points, as in Fig. 3.2 where contours and surfaces are plotted.

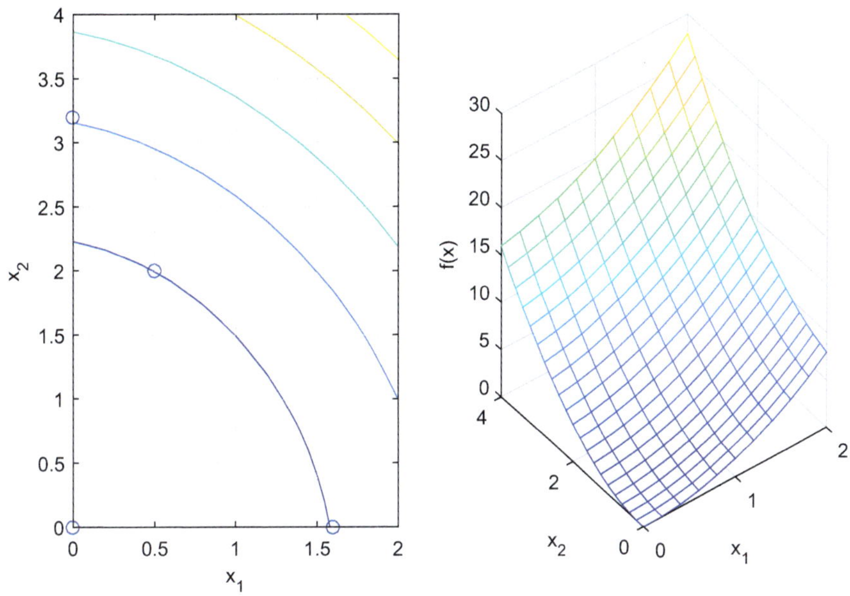

Fig. 3.2 Contour and surface plots of $f(\mathbf{x}) = 2x_1^2 + 0.5x_1x_2 + x_2^2$

(i) $\mathbf{x} = \begin{bmatrix} 0.5 \\ 2 \end{bmatrix}$. This is an interior point, and any feasible direction at this point is in the form of $\mathbf{d} = \begin{bmatrix} d_1 \\ d_2 \end{bmatrix}$, $d_1, d_2 \in \mathbb{R}$. Therefore, $\mathbf{d}^T \nabla f(\mathbf{x}) = 3d_1 + 4.25d_2$ which is not zero in general and thus does not satisfy the condition.

(ii) $\mathbf{x} = \begin{bmatrix} 1.6 \\ 0 \end{bmatrix}$. This is a boundary point, and any feasible direction at this point is in the form of $\mathbf{d} = \begin{bmatrix} d_1 \\ d_2 \end{bmatrix}$, $d_1 \in \mathbb{R}, d_2 > 0$. So, $\mathbf{d}^T \nabla f(\mathbf{x}) = 6.4d_1 + 0.8d_2$ which can take both positive and negative values. Therefore, it does not satisfy the condition.

(iii) $\mathbf{x} = \begin{bmatrix} 0 \\ 3.2 \end{bmatrix}$. This is a boundary point, and any feasible direction at this point is in the form of $\mathbf{d} = \begin{bmatrix} d_1 \\ d_2 \end{bmatrix}$, $d_1 > 0, d_2 \in \mathbb{R}$. Therefore, $\mathbf{d}^T \nabla f(\mathbf{x}) = 1.6d_1 + 6.4d_2$ that can take both positive and negative values and thus does not satisfy the condition.

(iv) $\mathbf{x} = \begin{bmatrix} 0 \\ 0 \end{bmatrix}$. This is also a boundary point, and any feasible direction at this point is in the form of $\mathbf{d} = \begin{bmatrix} d_1 \\ d_2 \end{bmatrix}$, $d_1 > 0, d_2 > 0$. Since $\mathbf{d}^T \nabla f(\mathbf{x}) = 0$, it satisfies the first-order necessary condition. □

As another necessary condition, the second-order necessary condition, utilizing the second-order derivative or the Hessian function, is stated in Theorem 2.

Theorem 2 (Second-Order Necessary Condition) *Suppose that* $f : \Omega \to \mathbb{R}$ *is twice differentiable over* Ω *and* $\mathbf{d}^T \nabla f(\mathbf{x}^*) = 0$ *for all feasible directions* $\mathbf{d}$ *at* $\mathbf{x}^*$. *Moreover, suppose that* $\mathbf{H}(\mathbf{x}^*)$ *is the Hessian of* f *at* $\mathbf{x}^*$.

(a) If $\mathbf{x}^*$ *is a local minimizer of* f *over* Ω, *then* $\mathbf{d}^T \mathbf{H}(\mathbf{x}^*)\mathbf{d} \geq 0$ *for all feasible directions* $\mathbf{d}$.
(b) If $\mathbf{x}^*$ *is a local maximizer of* f *over* Ω, *then* $\mathbf{d}^T \mathbf{H}(\mathbf{x}^*)\mathbf{d} \leq 0$ *for all feasible directions* $\mathbf{d}$.

Proof The proof is given for part (a). Part (b) follows by considering function $-f$. Based on the Taylor series, the second-order approximation of f at $\mathbf{x}^*$ is given by:

$$f(\mathbf{x}^* + \alpha \mathbf{d}) \approx f(\mathbf{x}^*) + \alpha \mathbf{d}^T \nabla f(\mathbf{x}^*) + \frac{\alpha^2}{2} \mathbf{d}^T \mathbf{H}(\mathbf{x}^*)\mathbf{d}.$$

If $\mathbf{d}^T \nabla f(\mathbf{x}^*) = 0$ and $f(\mathbf{x}^* + \alpha \mathbf{d}) \geq f(\mathbf{x}^*)$, then $\mathbf{d}^T \mathbf{H}(\mathbf{x}^*)\mathbf{d} \geq 0$. □

As a special case in Theorem 2, if $\mathbf{x}^*$ is an optimal interior point, $\mathbf{d}^T \mathbf{H}(\mathbf{x}^*)\mathbf{d} \geq 0$ for all $\mathbf{d} \in \mathbb{R}^n$, as the whole of $\mathbb{R}^n$ is feasible direction. This implies that $\mathbf{H}(\mathbf{x}^*)$ should be a positive semi-definite matrix. Similarly, for the case of local maximizer, $\mathbf{H}(\mathbf{x}^*)$ should be negative semi- definite.

Note that the aforementioned necessary conditions in Theorems 1 and 2 are not sufficient for optimal solution. For example, for $f(x) = x^3$, we derive $f'(x) = 3x^2$, and $f''(x) = 6x$. The point $x = 0$ satisfies the first-order and second-order necessary conditions. However, this point is neither local minimum nor local maximum. Indeed, the points which satisfy the necessary conditions are only candidates to be investigated for optimal solutions.

Example 2 Find the points which satisfy the necessary conditions of optimality for $f(x) = x_1^2 + x_2^2$, $\mathbf{x} \in \mathbb{R}^2$.

Solution This is an unconstrained optimization problem:

$$\nabla f(\mathbf{x}) = \begin{bmatrix} 2x_1 \\ 2x_2 \end{bmatrix} = \mathbf{0} \Rightarrow \mathbf{x}^* = \begin{bmatrix} 0 \\ 0 \end{bmatrix} \Rightarrow \mathbf{H}(\mathbf{x}^*) = \begin{bmatrix} 2 & 0 \\ 0 & 2 \end{bmatrix}$$

which is a positive semi-definite matrix. Therefore, $\mathbf{x}^* = \begin{bmatrix} 0 \\ 0 \end{bmatrix}$ satisfies the necessary condition of a local minimum point. □

Example 3 Find the points which satisfy the necessary conditions of optimality for $f(\mathbf{x}) = x_2^2$, $\mathbf{x} \in \mathbb{R}^2$.

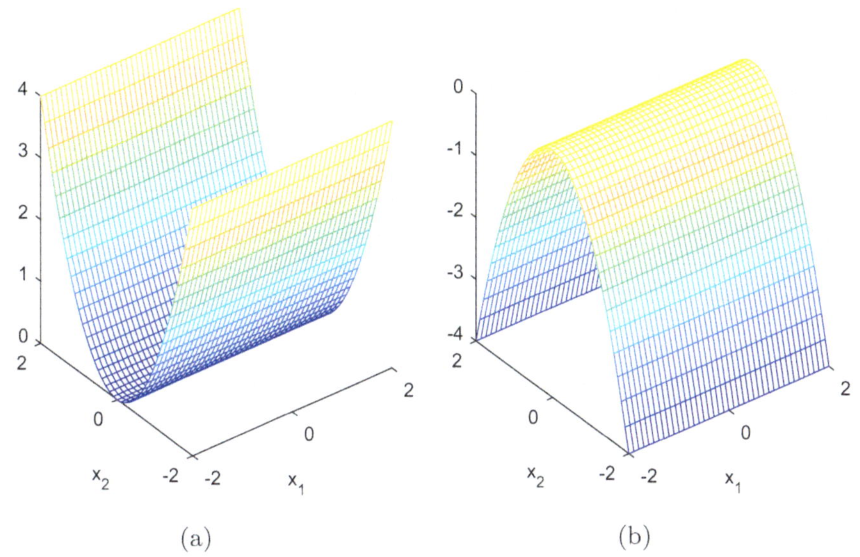

Fig. 3.3 Plots of **a**) $f(\mathbf{x}) = x_2^2$ and **b**) $f(\mathbf{x}) = -x_2^2$

Solution This is an unconstrained problem:

$$\nabla f(\mathbf{x}) = \begin{bmatrix} 0 \\ 2x_2 \end{bmatrix} = 0 \Rightarrow x^* = \begin{bmatrix} x_1 \\ 0 \end{bmatrix}, x_1 \in \mathbb{R} \Rightarrow \mathbf{H}(\mathbf{x}^*) = \begin{bmatrix} 0 & 0 \\ 0 & 2 \end{bmatrix}$$

which is a positive semi-definite matrix. Therefore, any point in the form of $\mathbf{x}^* = \begin{bmatrix} x_1 \\ 0 \end{bmatrix}$ satisfies the necessary conditions of a local minimum point. Similarly, taking the same operation, any point in the form of $\mathbf{x}^* = \begin{bmatrix} x_1 \\ 0 \end{bmatrix}$ satisfies the necessary conditions of a local maximum point of $f(\mathbf{x}) = -x_2^2$, $\mathbf{x} \in \mathbb{R}^2$. These observations are plotted in Fig. 3.3. □

Example 4 Find the points which satisfy the necessary conditions of optimality for $f(x) = x_1^2 - x_2^2$, $\mathbf{x} \in \mathbb{R}^2$.

Solution Consider

$$\nabla f(x) = \begin{bmatrix} 2x_1 \\ -2x_2 \end{bmatrix} = 0 \Rightarrow \mathbf{x}^* = \begin{bmatrix} 0 \\ 0 \end{bmatrix} \Rightarrow \mathbf{H}(\mathbf{x}^*) = \begin{bmatrix} 2 & 0 \\ 0 & -2 \end{bmatrix}$$

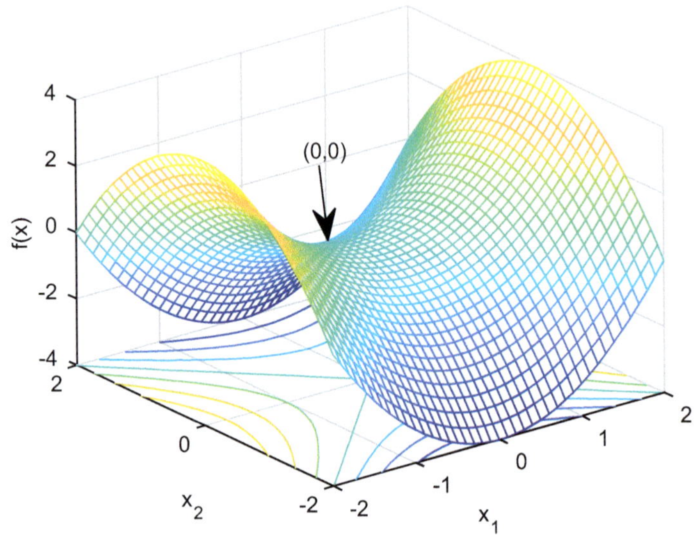

Fig. 3.4 Illustration of saddle point for $f(x) = x_1^2 - x_2^2$

that is indefinite. For any feasible direction $\mathbf{d} = \begin{bmatrix} d_1 \\ d_2 \end{bmatrix}$, we have $\mathbf{d}^T\mathbf{H}(\mathbf{x}^*)\mathbf{d} = 2d_1^2 - 2d_2^2$, which can be either positive or negative depending on $\mathbf{d}$. Therefore, $\mathbf{x}^* = \begin{bmatrix} 0 \\ 0 \end{bmatrix}$ is neither local minimum nor local maximum. In particular, this point is called a *saddle* point, as shown in Fig. 3.4. □

So far, necessary conditions of optimality have been discussed. In general, it is not possible to state a sufficient condition for optimal solutions. However, in the case of interior points, this condition is stated in Theorem 3.

Theorem 3 (Second-Order Sufficient Condition) *Suppose that $f : \Omega \to \mathbb{R}$ is twice differentiable over Ω, $\mathbf{x}^*$ is an interior point and $\nabla f(\mathbf{x}^*) = 0$.*

(a) If $\mathbf{H}(\mathbf{x}^*) > 0$, then $\mathbf{x}^*$ is a strict local minimizer of f over Ω.
(b) If $\mathbf{H}(\mathbf{x}^*) < 0$, then $\mathbf{x}^*$ is a strict local maximizer of f over Ω.

Proof The proof is given for part (a). Part (b) follows by considering function $-f$. As stated, based on the Taylor series, the second-order approximation of f at $\mathbf{x}^*$ is

$$f(\mathbf{x}^* + \alpha\mathbf{d}) \approx f(\mathbf{x}^*) + \alpha\mathbf{d}^T\nabla f(\mathbf{x}^*) + \frac{\alpha^2}{2}\mathbf{d}^T\mathbf{H}(\mathbf{x}^*)\mathbf{d}.$$

If $\nabla f(\mathbf{x}^*) = 0$ and $\mathbf{d}^T\mathbf{H}(\mathbf{x}^*)\mathbf{d} > 0$ for all $\mathbf{d}$, we derive $f(\mathbf{x}^*) < f(\mathbf{x}^* + \alpha\mathbf{d})$. Therefore, $\mathbf{x}^*$ is a strict local minimizer. □

Example 5 Find local minimum and maximum points of $f(\mathbf{x}) = x_1 e^{-(x_1^2+x_2^2)}$, $\mathbf{x} \in \mathbb{R}^2$.

Solution To check out the sufficient conditions, consider

$$\nabla f(\mathbf{x}) = e^{-(x_1^2+x_2^2)} \begin{bmatrix} 1-2x_1^2 \\ -2x_1x_2 \end{bmatrix} = 0 \Rightarrow \mathbf{x}_1^* = \begin{bmatrix} 0.7 \\ 0 \end{bmatrix}, \ \mathbf{x}_2^* = \begin{bmatrix} -0.7 \\ 0 \end{bmatrix}$$

and

$$\nabla^2 f(\mathbf{x}) = e^{-(x_1^2+x_2^2)} \begin{bmatrix} 4x_1^2-6x_1 & 4x_1^2x_2-2x_2 \\ 4x_1^2x_2-2x_2 & 4x_1x_2^2-2x_1 \end{bmatrix}.$$

(i) $\nabla^2 f(\mathbf{x}_1^*) = \begin{bmatrix} -1.3 & 0 \\ 0 & -0.85 \end{bmatrix} < 0 \Rightarrow \mathbf{x}_1^*$ is a strict local maximum.

(ii) $\nabla^2 f(\mathbf{x}_2^*) = \begin{bmatrix} 3.7 & 0 \\ 0 & 0.85 \end{bmatrix} > 0 \Rightarrow \mathbf{x}_2^*$ is a strict local minimum.

This derivation is observed in Fig. 3.5. □

Example 6 Find local minimum and maximum points of $f(\mathbf{x}) = 2x_1^3 + 3x_2^2 + 3x_1^2x_2 - 24x_2$, $\mathbf{x} \in \mathbb{R}^2$.

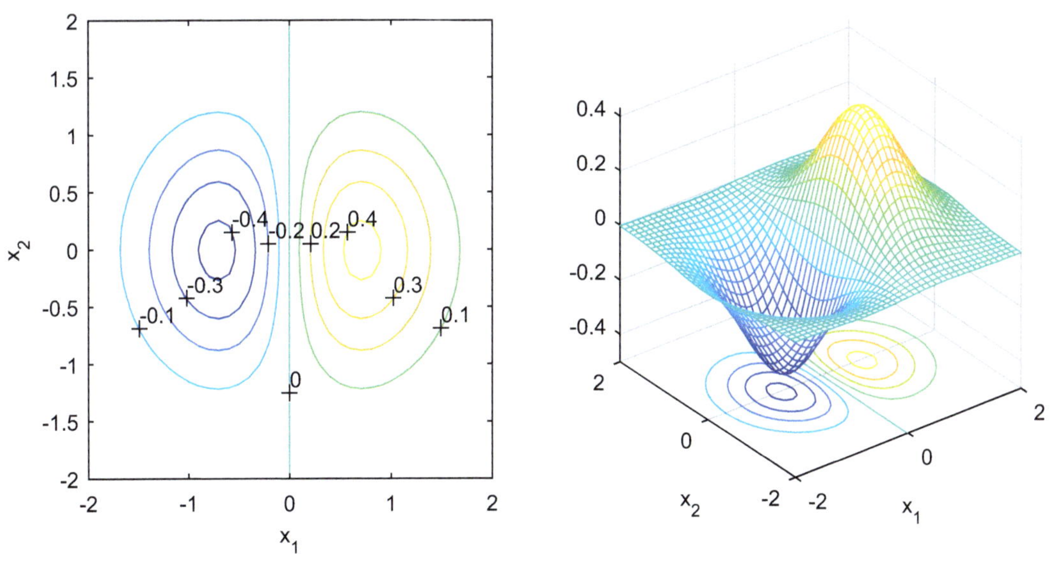

Fig. 3.5 Contour and surface plots of $f(\mathbf{x}) = x_1 e^{-(x_1^2+x_2^2)}$

Solution To check out the optimality conditions, consider

$$\nabla f(\mathbf{x}) = \begin{bmatrix} 6x_1^2 + 6x_1x_2 \\ 6x_2 + 3x_1^2 - 24 \end{bmatrix} = 0$$

$$\Rightarrow \mathbf{x}_1^* = \begin{bmatrix} 0 \\ 4 \end{bmatrix}, \mathbf{x}_2^* = \begin{bmatrix} 4 \\ -4 \end{bmatrix}, \mathbf{x}_3^* = \begin{bmatrix} -2 \\ 2 \end{bmatrix}$$

Then, we check if:

$$H(\mathbf{x}) = \begin{bmatrix} 12x_1 + 6x_2 & 6x_1 \\ 6x_1 & 6 \end{bmatrix}$$

is positive or negative definite in the derived points.

i) $H(\mathbf{x}_1^*) = \begin{bmatrix} 24 & 0 \\ 0 & 6 \end{bmatrix}$. It is a diagonal matrix with positive entries in the diagonal. Therefore, $H(\mathbf{x}_1^*) > 0$ and consequently $\mathbf{x}_1^*$ is a strict local minimum point.

ii) $H(\mathbf{x}_2^*) = \begin{bmatrix} 24 & 24 \\ 24 & 6 \end{bmatrix}$. This symmetric matrix has both positive and negative eigenvalues. Therefore, $H(\mathbf{x}_2^*)$ is indefinite and $\mathbf{x}_2^*$ is a saddle point.

iii) $H(\mathbf{x}_3^*) = \begin{bmatrix} -12 & -12 \\ -12 & 6 \end{bmatrix}$. This symmetric matrix has both positive and negative entries in the diagonal. Therefore, $H(\mathbf{x}_3^*)$ is indefinite and $\mathbf{x}_3^*$ is also a saddle point.

□

3.4 Least Square Optimization

Consider the set of linear equations

$$\mathbf{Ax} = \mathbf{b} \tag{3.4}$$

where $\mathbf{A} \in \mathbb{R}^{m \times n}$, $\mathbf{x} \in \mathbb{R}^n$, $\mathbf{b} \in \mathbb{R}^m$, and rank $\mathbf{A} = n$. If $\mathbf{b}$ is in the range space of $\mathbf{A}$ and $\mathbf{A}$ is nonsingular, then the unique solution is $\mathbf{x}^* = \mathbf{A}^{-1}\mathbf{b}$. However, suppose that $\mathbf{b}$ is not in the range space of $\mathbf{A}$. Therefore, the equation is said to be inconsistent and it has no solution. In this case, instead of searching to find the solution of $\mathbf{Ax} = \mathbf{b}$, it is of interest to find a point $\mathbf{x}^*$ such that $\|A\mathbf{x}^* - \mathbf{b}\|^2 \leq \|A\mathbf{x} - \mathbf{b}\|^2$ for all $\mathbf{x} \in \mathbb{R}^n$. In other words, the linear equation solution reduces to problem:

$$\min_{\mathbf{x}} \quad \|A\mathbf{x} - \mathbf{b}\|^2. \tag{3.5}$$

This problem is called the least square optimization problem, and the optimal point $\mathbf{x}^*$ is called the least square solution. The solution of this problem is discussed in Theorem 4.

Theorem 4 *The unique solution that minimizes* $\|A\mathbf{x} - \mathbf{b}\|^2$ *is given as* $\mathbf{x}^* = (\mathbf{A}^T\mathbf{A})^{-1}\mathbf{A}^T\mathbf{b}$.

Proof Let's write the objective function as:

$$\begin{aligned} f(\mathbf{x}) &= \|A\mathbf{x} - \mathbf{b}\|^2 = (A\mathbf{x} - \mathbf{b})^T(A\mathbf{x} - \mathbf{b}) \\ &= \mathbf{x}^T\mathbf{A}^T\mathbf{A}\mathbf{x} - 2\mathbf{b}^T\mathbf{A}\mathbf{x} + \mathbf{b}^T\mathbf{b}. \end{aligned}$$

Taking the gradient, we derive

$$\nabla f(\mathbf{x}) = 0 \Rightarrow 2\mathbf{A}^T\mathbf{A}\mathbf{x} - 2\mathbf{A}^T\mathbf{b} = 0 \Rightarrow$$
$$\mathbf{x}^* = (\mathbf{A}^T\mathbf{A})^{-1}\mathbf{A}^T\mathbf{b}.$$

Moreover, $H(\mathbf{x}^*) = 2\mathbf{A}^T\mathbf{A}$, which is always positive definite. The reader can verify this statement. Therefore, $\mathbf{x}^* = (\mathbf{A}^T\mathbf{A})^{-1}\mathbf{A}^T\mathbf{b}$ satisfies the necessary and sufficient conditions of optimality and it is the minimum point. □

Example 7 Consider a wireless communication channel in Fig. 3.6 with transmit power p_t. The received power p_r in decibel (dB) is derived using the model:

$$p_r = p_t + K - 10\gamma \log_{10} d$$

where K is a constant depending on the radio frequency and antennas gains, γ is the path loss exponent, and d in meters is the distance between the transmitter and the receiver. In a set of empirical measurements of $p_r - p_t$ in dB, given in Table 3.1, find constants K and γ to minimize the mean square error between the model and the empirical measurements.

Solution Let's define $M_{\text{model}} \triangleq p_r - p_t = K - 10\gamma \log_{10} d$ and $M_{\text{measure}} = p_r - p_t$ given in Table 3.1. The mean square error is

$$e = \frac{1}{5}\sum_{i=1}^{5}(M_{\text{model}}(d_i) - M_{\text{measure}}(d_i))^2$$
$$= \frac{1}{5}\|\mathbf{A}\mathbf{x} - \mathbf{b}\|^2$$

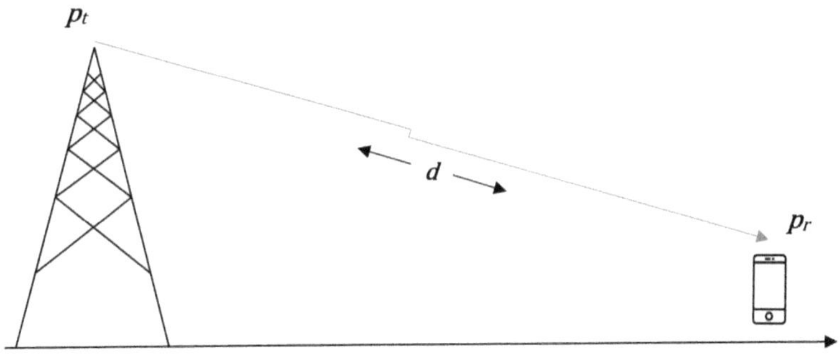

Fig. 3.6 Wireless communication channel

Table 3.1 Empirical measurements

d	$p_r - p_t$
10 m	−70 dB
20 m	−75 dB
50 m	−90 dB
100 m	−110 dB
300 m	−125 dB

where

$$\mathbf{x} = \begin{bmatrix} K \\ \gamma \end{bmatrix}, \ \mathbf{A} = \begin{bmatrix} 1 & -10\log_{10} 10 \\ 1 & -10\log_{10} 20 \\ 1 & -10\log_{10} 50 \\ 1 & -10\log_{10} 100 \\ 1 & -10\log_{10} 300 \end{bmatrix}, \ \mathbf{b} = \begin{bmatrix} -70 \\ -75 \\ -90 \\ -110 \\ -125 \end{bmatrix}.$$

Substituting in $\mathbf{x}^* = (\mathbf{A}^T\mathbf{A})^{-1}\mathbf{A}^T\mathbf{b}$, we derive $\mathbf{x}^* = \begin{bmatrix} -26.74 \\ 3.96 \end{bmatrix}$ and thus $K = -26.74$ and $\gamma = 3.96$. Moreover, the following code in MATLAB yields the same solution:

```
>> A=[1 -10*log10(10); 1 -10*log10(20); 1 -10*log10(50); ...
1 -10*log10(100);1 -10*log10(300)];
>> b=[-70;-75;-90;-110;-125];
>> x = A\b

x =

  -26.7440
    3.9669
```

□

3.5 General Optimization Algorithms

So far, we have demonstrated necessary and sufficient conditions for optimal solutions of problems in the form of:

$$\min_{\mathbf{x}} \ f(\mathbf{x}) \tag{3.6a}$$

$$\text{subject to } \ \mathbf{x} \in \Omega. \tag{3.6b}$$

Solving this problem analytically using the given conditions is almost difficult or even impossible in general. As alternative approaches, numerical and iterative algorithms are introduced to solve these kind of problems [3]. These algorithms typically require the user to supply a starting point $\mathbf{x}^0 \in \Omega$ as an initial solution candidate. Then, the algorithm generates a sequence of solutions $\{\mathbf{x}^k\}_{k=0}^{\infty}$ called iterates.

Under the assumption of given point $\mathbf{x}^k$, generating the next iterate $\mathbf{x}^{k+1}$ is based on a directional search method, which requires information on $\mathbf{x}^k$, $f(\mathbf{x}^k)$, and sometimes past points $\mathbf{x}^0, \ldots, \mathbf{x}^{k-1}$. The user is also required to supply the algorithm with a termination criterion to stop generating iterates. This criterion is to indicate that either the optimal solution has been reached within a desired accuracy or no further progress can be made any more. The aforementioned discussion is summarized as the following process:

1. **Initialization:** Supply a starting point $\mathbf{x}^0$ and a stopping criteria.
2. **Iteration** k: If $\mathbf{x}^k$ satisfies the termination criteria, then stop the algorithm. Otherwise:
   (a) Determine a search direction $\mathbf{d}^k$.
   (b) Determine a step size $\alpha^k > 0$.
   (c) Determine the next iterate $\mathbf{x}^{k+1} = \mathbf{x}^k + \alpha^k \mathbf{d}^k$.

More explanation on these steps and parameters are given in the following subsections.

3.5.1 Search Direction

Search direction $\mathbf{d}^k$ in the directional search methods is of high importance. In minimization problems, the search direction requires to have descent property, that is:

$$f(\mathbf{x}^k + \alpha^k \mathbf{d}^k) < f(\mathbf{x}^k). \tag{3.7}$$

Recall that $\nabla f(\mathbf{x}^k)^T \mathbf{d}^k$ is the rate of increase of f at $\mathbf{x}^k$. To ensure the descent property, we need to choose $\mathbf{d}^k$ such that $\nabla f(\mathbf{x}^k)^T \mathbf{d}^k < 0$. One form of $\mathbf{d}^k$ to satisfy this requirement is to choose

$$\mathbf{d}^k = -\mathbf{D}^k \nabla f(\mathbf{x}^k) \tag{3.8}$$

where $\mathbf{D}^k$ is a positive definite matrix. The reader can verify this statement. Based on the choice of $\mathbf{D}^k$, two well-known directional search methods are derived [3].

1. Steepest descent method with $\mathbf{D}^k = \mathbf{I}$.
2. Newton's method with $\mathbf{D}^k = \mathbf{H}(\mathbf{x}^k)^{-1}$, where $\mathbf{H}(\mathbf{x}^k)$ is the Hessian matrix and positive definite.

In the rest of this chapter, we only focus on these two methods.

3.5.2 Step Size

Having the point $\mathbf{x}^k$ and search direction $\mathbf{d}^k$ at each iterate k, the next important parameter in directional search methods is the step size α^k. It would be possible to either take small values or large values of step sizes. The first option reduces the speed of reaching the optimal solution, whereas the second option may result in a more zigzag path to this solution. Moreover, the second option may require fewer evaluations of search directions, because it may lead to divergence and instability. Therefore, it would be necessary to provide a trade-off between speed and accuracy.

On how to compute the step size, there exist two approaches. One simple but not fast approach is to choose a sufficiently small fixed value for the step size in all iterations, that is, $\alpha^k = \alpha =$ constant. Another approach is to choose α^k by solving the following one-dimensional optimization problem:

$$\min_{\alpha > 0} \quad \phi(\alpha) = f(\mathbf{x}^k + \alpha \mathbf{d}^k). \tag{3.9}$$

This is a one-dimensional problem and can be solved using $\phi'(\alpha) = 0$ or iterative methods such as the line search method.

3.5.3 Termination Criterion

As noted, the termination criterion is to ensure that optimality conditions have been reached or no further progress can be made in more iterations. Moreover, recall that interior point $\mathbf{x}^k$ is a local minimizer if and only if $\nabla f(\mathbf{x}^k) = 0$ and $\mathbf{H}(\mathbf{x}^k)$ is positive definite. These conditions can be used as bases for termination criterion. Satisfying these conditions in general requires high numerical computations and might not be fully achieved in a finite number of iterations. Therefore, it is necessary to provide a trade-off between accuracy and speed of convergence to the optimal solution. Following are a number of termination criteria, in which case we stop the algorithm:

1. One practical criterion is to check if the norm $\|\nabla f(\mathbf{x}^k)\|$ is less than a specified threshold ϵ, i.e., $\|\nabla f(\mathbf{x}^k)\| < \epsilon$.
2. Alternatively, the absolute difference between two successive iterations as $\|\mathbf{x}^{k+1} - \mathbf{x}^k\| < \epsilon$ or between their function values as $|f(\mathbf{x}^{k+1}) - f(\mathbf{x}^k)| < \epsilon$ can be used.
3. Finally, we may check relative values or scaled independent differences between two successive iterations as $\frac{\|\mathbf{x}^{k+1}-\mathbf{x}^k\|}{\max\{1,\|\mathbf{x}^k\|\}} < \epsilon$ or $\frac{|f(\mathbf{x}^{k+1})-f(\mathbf{x}^k)|}{\max\{1,|f(\mathbf{x}^k)|\}} < \epsilon$.

Having stated the general structure of optimization algorithms, following two well-known optimization algorithms including steepest descent method and Newton's method are presented.

3.6 Steepest Descent Method

Recall that $-\nabla f(\mathbf{x}^k)$ is the direction of the maximum decrease of f at point $\mathbf{x}^k$. Therefore, the search directional method using $\mathbf{d}^k = -\nabla f(\mathbf{x}^k)$ is called steepest descent method. Here, when a candidate solution $\mathbf{x}^k$ is given, the next iterate is generated as:

$$\mathbf{x}^{k+1} = \mathbf{x}^k - \alpha^k \nabla f(\mathbf{x}^k) \tag{3.10}$$

where $\alpha^k > 0$ is a positive scalar step size and is chosen such that:

$$\alpha^k = \arg\min_{\alpha \geq 0} f(\mathbf{x}^k - \alpha \nabla f(\mathbf{x}^k)). \tag{3.11}$$

In other words, α^k is chosen to minimize $f(\mathbf{x}^{k+1})$.

One important characteristic of steepest descent method is its descent property, i.e., $f(\mathbf{x}^{k+1}) < f(\mathbf{x}^k)$. To show this property, let's rewrite $f(\mathbf{x}^{k+1})$ as:

$$\begin{aligned} f(\mathbf{x}^{k+1}) = f(\mathbf{x}^k - \alpha^k \nabla f(\mathbf{x}^k)) &= f(\mathbf{x}^k) - \alpha^k \nabla f(\mathbf{x}^k)^T \nabla f(\mathbf{x}^k) + o(\alpha^k) \\ &= f(\mathbf{x}^k) - \alpha^k \|\nabla f(\mathbf{x}^k)\|^2 + o(\alpha^k). \end{aligned}$$

If $\nabla f(\mathbf{x}^k) \neq 0 \Rightarrow f(\mathbf{x}^{k+1}) < f(\mathbf{x}^k)$, as $o(\alpha^k)$ is sufficiently small. Therefore, $\mathbf{x}^{k+1}$ is an improvement over $\mathbf{x}^k$ if we are searching for a minimizer.

Another important characteristic of the steepest descent method is on the path of reaching from the initial point to the optimal solution. Theorem 5 states that this is a zigzag path with orthogonal steps.

Theorem 5 *In the steepest descent method, every two consecutive search directions are orthogonal to each other. In other words,* $\mathbf{x}^{k+1} - \mathbf{x}^k$ *is orthogonal* $\mathbf{x}^{k+2} - \mathbf{x}^{k+1}$.

Proof Consider the inner product $\langle \mathbf{x}^{k+1} - \mathbf{x}^k, \mathbf{x}^{k+2} - \mathbf{x}^{k+1} \rangle = \alpha^k \alpha^{k+1} \langle \nabla f(\mathbf{x}^k), \nabla f(\mathbf{x}^{k+1}) \rangle$. We also have $\alpha^k = \arg\min_{\alpha \geq 0} f(\mathbf{x}^k - \alpha \nabla f(\mathbf{x}^k)) = \arg\min_{\alpha \geq 0} \phi(\alpha)$. Therefore, $\phi'(\alpha^k) = 0 \Rightarrow \nabla f(\mathbf{x}^k)^T(-\nabla f(\mathbf{x}^{k+1})) = 0$ using the chain rule. This completes the proof. □

As a conclusion, steepest descent method is mostly simple to implement and thus is widely used in practical applications.

Example 8 Find the minimizer of $f(\mathbf{x}) = x_1 e^{-(x_1^2 + x_2^2)}$.

Solution This is the same function as given in Example 5. Using the derived gradient, the following code in MATLAB implements the steepest decent method for this example. The step size is optimally computed at each step. Results are given in Fig. 3.7a, where the first part is the contour plot and the second part is the function value starting from an initial point $\mathbf{x}^0$ to the optimal point $\mathbf{x}^* = \begin{bmatrix} -0.7 \\ 0 \end{bmatrix}$.

□

```
%================================================================
clear all , close all
Q = [1 0; 0 1];
f = @(x) x(1)*exp(-x'*Q*x);
g = @(x) exp(-x'*Q*x)*[1-2*x(1)^2; -2*x(1)*x(2) ];
delta = 0.1;
xx=-1.6:delta:0; yy=-1.6:delta:1.6;
[X, Y] = meshgrid(xx, yy);
Z = X.*exp(-X.^2 - Y.^2);
stepsize = 0.25;
epsi = 0.001;
k=1;
x(:,k) = [-1; 1];
fvalue(k) = f(x(:,k));
for k=1:1000
    a = g(x(:, k));
    xc= x(1,k);    yc=x(2,k);
    phi = @(s) (xc-s*a(1))*exp(-(xc-s*a(1))^2 - (yc-s*a(2))^2);
    stepsize = fminbnd(phi, 0, 10);
    x(:,k+1) = x(:,k) - stepsize*g(x(:,k));
    fvalue(k+1) = f(x(:,k+1));
    tol = abs(f(x(:,k+1)) - f(x(:,k)))/max(1, abs(f(x(:,k))));
    if tol<epsi
        break;
    end
end
figure(1), subplot(1,2,1), contour(X,Y,Z,fvalue);
hold on, plot(x(1,:), x(2,:), 'o-'), xlabel('x_1'), ylabel('x_2')
subplot(1,2,2), semilogy(1:k+1, fvalue, '-o'), xlabel('k'),...
ylabel('f(x)')
%================================================================
```

To demonstrate the effectiveness of optimal step sizes, the steepest decent method with fixed step size $\alpha^k = \alpha = 0.25$ is also done. The results are depicted in Fig. 3.7a. As shown, in comparison to optimal step sizes in Fig. 3.7b, more steps have been taken to reach the optimal solution. It is noteworthy that, depending on the starting point, search methods may converge to a local minimum.

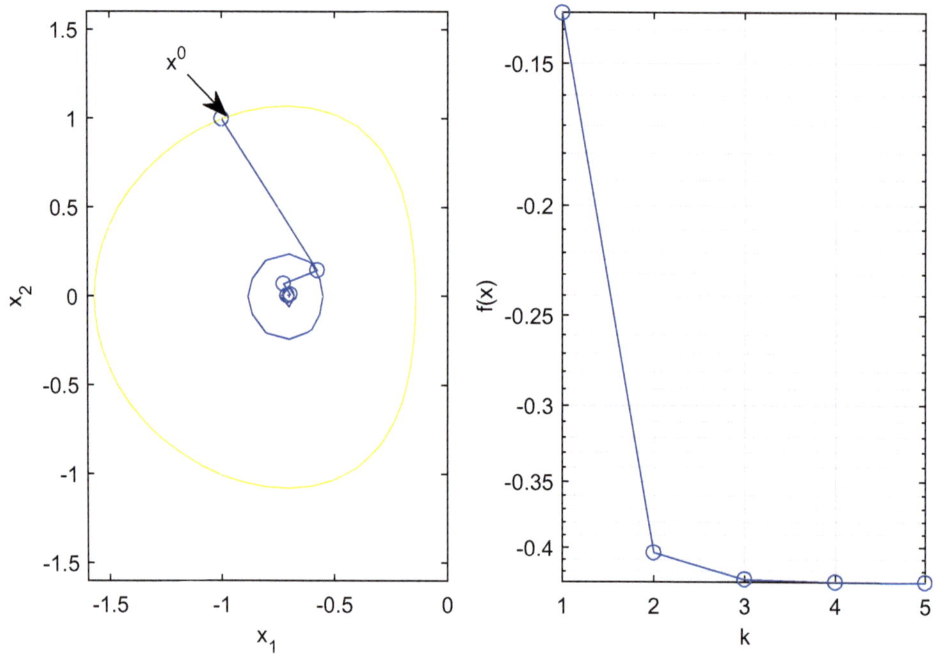

(a)

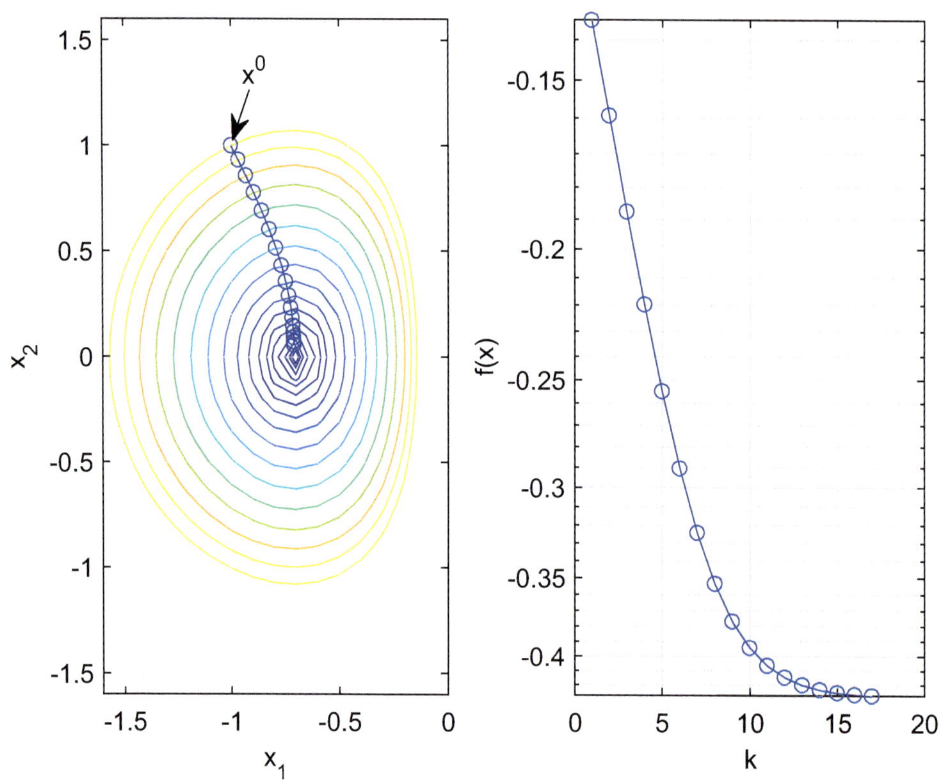

(b)

Fig. 3.7 Contour plots and function values versus iterations with $\epsilon = 0.001$ for $f(\mathbf{x}) = x_1 e^{-(x_1^2+x_2^2)}$. (**a**) Optimal step sizes α^k. (**b**) Fixed step size $\alpha = 0.25$

Example 9 Find the minimizer of $f(\mathbf{x}) = 0.06e^{2x_1+x_2} + 0.05e^{x_1-2x_2} + e^{-x_1}$.

Solution The gradient is

$$\nabla f(\mathbf{x}) = \begin{bmatrix} 0.12e^{2x_1+x_2} + 0.05e^{x_1-2x_2} - e^{-x_1} \\ 0.06e^{2x_1+x_2} - 0.1e^{x_1-2x_2} \end{bmatrix}$$

and the following code is the steepest method for this example. Results are given in Fig. 3.8a, where the first part is the contour plot and the second part is the function value starting from an initial point $\mathbf{x}^0$ to the optimal point $\mathbf{x}^* = \begin{bmatrix} 0.64 \\ -0.02 \end{bmatrix}$. The orthogonality of every two consecutive search directions is clearly shown in this figure. □

```
%=================================================================
clear all, close all
c1=0.06;   c2=0.05;   c3=1;
f = @(x) c1*exp(2*x(1) + x(2)) + c2*exp(x(1) - 2*x(2)) + c3*exp(-x(1));
g = @(x)  [2*c1*exp(2*x(1)+x(2)) + c2*exp(x(1) - 2*x(2)) ...
- c3 * exp(-x(1));  c1*exp(2*x(1) + x(2)) - 2*c2*exp(x(1)- 2*x(2))];
delta = 0.1;
xx=-1:delta:1.75; yy=-2:delta:3.5;
[X, Y] = meshgrid(xx, yy);
Z=c1*exp(2*X + Y ) + c2*exp(X - 2*Y ) + c3*exp(-X);
stepsize = 0.25;
epsi = 0.001;
k=1;
x(:,k) = [-0.5;2];
fvalue(k) = f(x(:,k));
for k=1:1000
    a = g(x(:, k));
    xc= x(1,k);     yc=x(2,k);
    phi = @(s) c1*exp(2*(xc - s*a(1)) + (yc-s*a(2))) ...
    c2*exp((xc - s*a(1))- 2*(yc-s*a(2))) +c3*exp(-(xc - s*a(1)));
    stepsize = fminbnd(phi, 0, 10);
    x(:,k+1) = x(:,k) - stepsize*g(x(:,k));
    fvalue(k+1) =  f(x(:,k+1));
    tol = abs(f(x(:,k+1)) - f(x(:,k)))/max(1, abs(f(x(:,k))));
    if tol<epsi
         break;
     end
 end
figure(1), subplot(1,2,1), contour(X,Y,Z,fvalue);
hold on, plot(x(1,:), x(2,:), 'o-'), xlabel('x_1'), ylabel('x_2')
subplot(1,2,2), semilogy(1:k+1, fvalue, '-o'), xlabel('k'),...
 ylabel('f(x)')
 %=================================================================
```

To demonstrate the effectiveness of optimal step sizes, the steepest decent method with constant step size $\alpha^k = \alpha = 0.25$ is also done. The results are given in Fig. 3.8b. As shown, in comparison with optimal step sizes in Fig. 3.8a, more steps have been taken to reach the optimal solution.

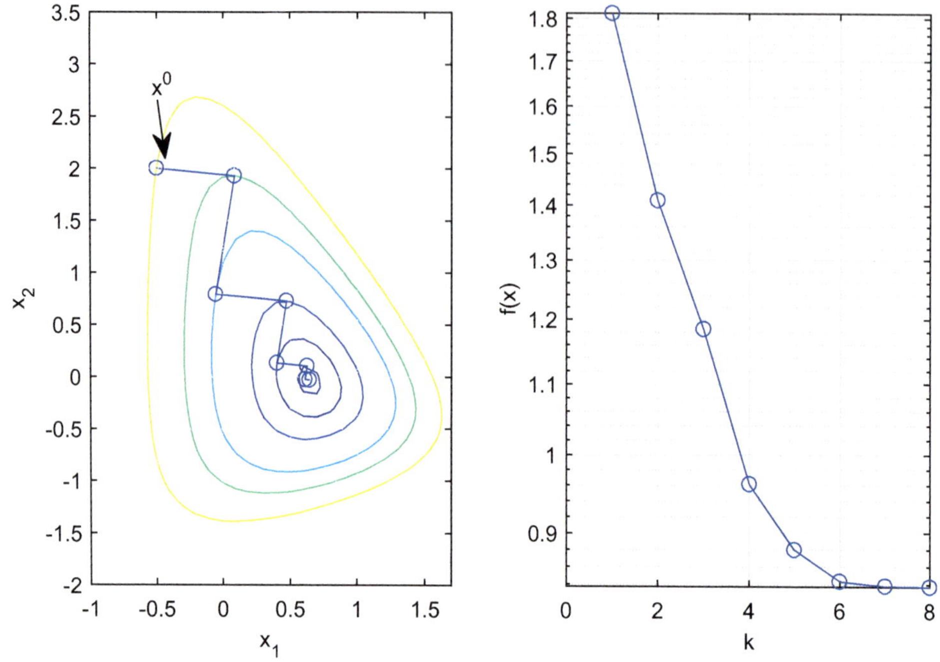

(a)

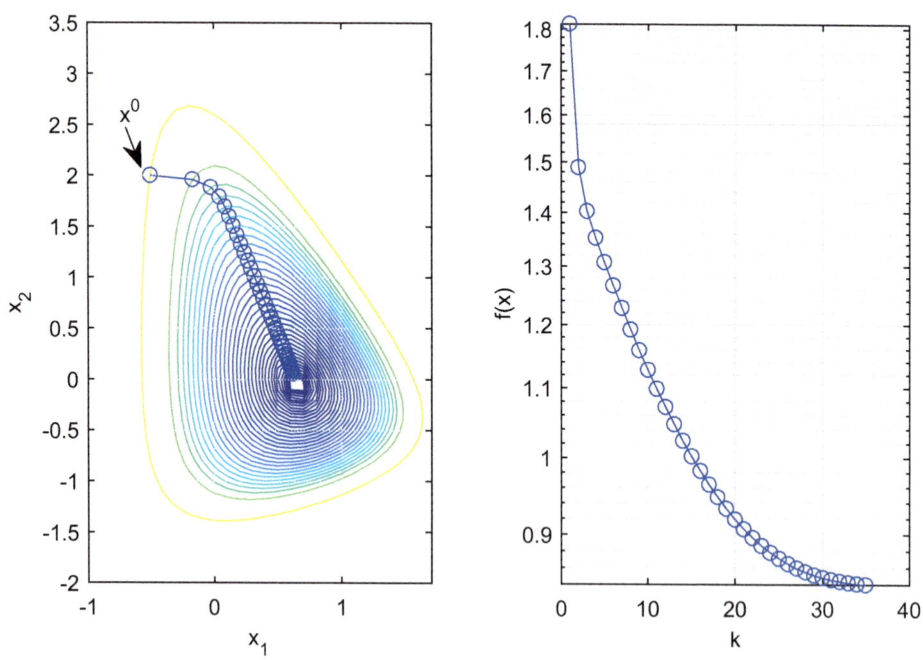

(b)

Fig. 3.8 Contour plots and function values versus iterations with $\epsilon = 0.001$ for $f(\mathbf{x}) = 0.06e^{2x_1+x_2} + 0.05e^{x_1-2x_2} + e^{-x_1}$. (**a**) Optimal step sizes α^k. (**b**) Constant step size $\alpha = 0.25$

3.7 Newton's Method

While the steepest descent method uses only the first-order derivative of the function in the search direction, the Newton's method uses both the first- and the second-order derivations of the function in the hope of performance improvement. The idea behind Newton's method is to construct a quadratic approximation of the function at each iteration such that the first- and the second-order derivatives of the function and its approximation matches with each other. The method then minimizes the derived quadratic function to generate the next iterate. As a special case, if $f(\mathbf{x})$ is quadratic, the optimal point is attained in only one step. Otherwise, the approximation will only provide an estimate of the minimizer.

To justify the search direction in this method, note that every twice differentiable function $f(\mathbf{x})$ can be approximated as a quadratic function in a close proximity of $\mathbf{x}^k$ as:

$$\begin{aligned} f(\mathbf{x}) \approx q(\mathbf{x}) &= f(\mathbf{x}^k) + \nabla f(\mathbf{x}^k)^T(\mathbf{x} - \mathbf{x}^k) \\ &+ \frac{1}{2}(\mathbf{x} - \mathbf{x}^k)^T \mathbf{H}(\mathbf{x}^k)(\mathbf{x} - \mathbf{x}^k) \end{aligned} \tag{3.12}$$

which is a quadratic Taylor expansion of f at $\mathbf{x}^k$. The minimum point of $q(\mathbf{x})$ is obtained by $\nabla q(\mathbf{x}) = 0$ or

$$\begin{aligned} &\nabla f(\mathbf{x}^k) + \mathbf{H}(\mathbf{x}^k)(\mathbf{x} - \mathbf{x}^k) = 0 \\ &\Rightarrow \mathbf{x} = \mathbf{x}^k - \mathbf{H}(\mathbf{x}^k)^{-1}\nabla f(\mathbf{x}^k). \end{aligned} \tag{3.13}$$

Therefore, the direction $\mathbf{d}^k = -\mathbf{H}(\mathbf{x}^k)^{-1}\nabla f(\mathbf{x}^k)$ can be considered as the search direction in the Newton's method. Thus, the iterative search method is formed as:

$$\mathbf{x}^{k+1} = \mathbf{x}^k - \mathbf{H}(\mathbf{x}^k)^{-1}\nabla f(\mathbf{x}^k). \tag{3.14}$$

Based on Theorem 6, the Newton's method is not always guaranteed to have descent property, unless $\mathbf{H}(\mathbf{x}^k)$ is positive definite.

Theorem 6 *If* $\mathbf{H}(\mathbf{x}^k)$ *is positive definite, then* $\mathbf{d}^k = -\mathbf{H}(\mathbf{x}^k)^{-1}\nabla f(\mathbf{x}^k)$ *is a descent direction.*

Proof Since $\mathbf{H}(\mathbf{x}^k)$ is positive definite, for any $\mathbf{v} \neq 0$, we have that

$$\begin{aligned} &(\mathbf{H}(\mathbf{x}^k)^{-1}\mathbf{v})^T \mathbf{H}(\mathbf{x}^k)(\mathbf{H}(\mathbf{x}^k)^{-1}\mathbf{v}) > 0 \\ &\Rightarrow \mathbf{v}^T \mathbf{H}(\mathbf{x}^k)^{-T}\mathbf{v} > 0. \end{aligned}$$

Therefore, $\mathbf{H}(\mathbf{x}^k)^{-T}$ and thus $\mathbf{H}(\mathbf{x}^k)^{-1}$ is also positive definite or equivalently $-\mathbf{H}(\mathbf{x}^k)^{-1}$ is negative definite. This implies that

$$-\nabla f(\mathbf{x}^k)^T \mathbf{H}(\mathbf{x}^k)^{-1}\nabla f(\mathbf{x}^k) < 0$$

and therefore $\mathbf{d}^k = -\mathbf{H}(\mathbf{x}^k)^{-1}\nabla f(\mathbf{x}^k)$ is a descent direction. □

Note that the Newton's method assumes that $\mathbf{H}(\mathbf{x}^k)$ is nonsingular to have the inverse. Moreover, the points generated by this method may not converge in general, even if $\mathbf{H}(\mathbf{x}^k)$ is nonsingular, unless the starting point is close enough to the optimal point. Furthermore, this method may converge to a local minimum.

Example 10 Find the minimizer of $f(\mathbf{x}) = -\ln(1-x_1-x_2) - \ln(x_1) - \ln(x_2)$ using Newton's method.

Solution Prior to use the Newton's method, we need to ensure that the Hessian is always positive definite. Consider

$$\nabla f(\mathbf{x}) = \begin{bmatrix} \frac{1}{1-x_1-x_2} - \frac{1}{x_1} \\ \frac{1}{1-x_1-x_2} - \frac{1}{x_2} \end{bmatrix}$$

and

$$\mathbf{H}(\mathbf{x}) = \begin{bmatrix} \left(\frac{1}{1-x_1-x_2}\right)^2 + \left(\frac{1}{x_1}\right)^2 & \left(\frac{1}{1-x_1-x_2}\right)^2 \\ \left(\frac{1}{1-x_1-x_2}\right)^2 & \left(\frac{1}{1-x_1-x_2}\right)^2 + \left(\frac{1}{x_2}\right)^2 \end{bmatrix}.$$

It can be verified that $\mathbf{H}(\mathbf{x}) > 0$ for all $\mathbf{x}$. The following code in MATLAB implements the Newton's method for this example. The results are shown in Fig. 3.9. The first part is the contour plot of $f(\mathbf{x})$ starting from an initial point $\mathbf{x}^0$ to the optimal point, and the second one is the function value along iterations k. After a number of iterations, optimal point is reached as $\mathbf{x}^* = \begin{bmatrix} 0.3084 \\ 0.3452 \end{bmatrix}$.

□

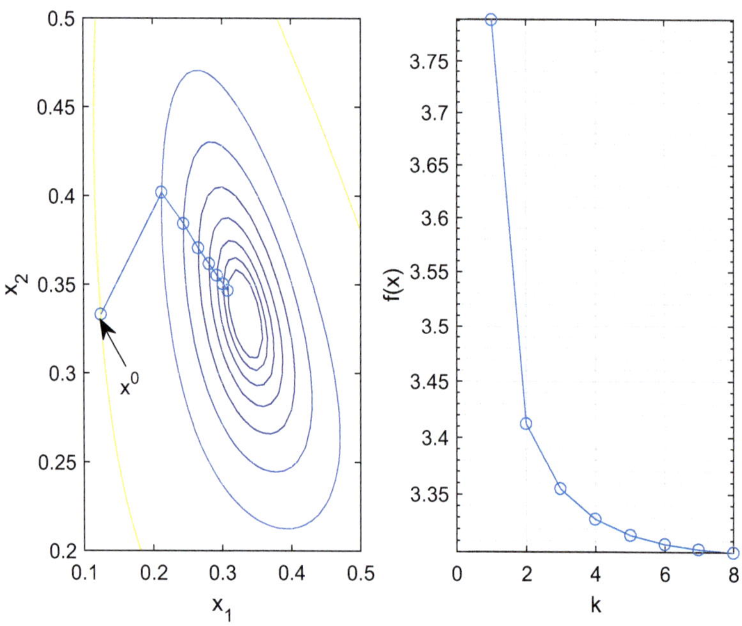

Fig. 3.9 Newton's method: contour plots and function values versus iterations for $f(\mathbf{x}) = -\ln(1-x_1-x_2) - \ln(x_1) - \ln(x_2)$

```
%====================================================================
clear all , close all
f = @(x) -log(1-x(1)-x(2)) - log(x(1)) - log(x(2));
g = @(x) [1/(1-x(1)-x(2))-1/x(1);  1/(1-x(1)-x(2))-1/x(2) ];
delta = 0.01;
xx=0.1:delta:0.5; yy=0.2:delta:0.5;
[X, Y] = meshgrid(xx, yy);
Z = -log(1-X-Y) - log(X)-log(Y);
epsi = 0.001;
k=1;
x(:,k) = [1/8; 1/3];
fvalue(k) = f(x(:,k));
for k=1:1000
   H  = [1/(1-x(1)-x(2))^2 + (1/x(1))^2   1/(1-x(1)-x(2))^2; ...
   1/(1-x(1)-x(2))^2    1/(1-x(1)-x(2))^2 + (1/x(2))^2];
   x(:,k+1) = x(:,k) - inv(H)*g(x(:,k));
   fvalue(k+1) =  f(x(:,k+1));
   tol = abs(f(x(:,k+1)) - f(x(:,k)))/max(1, abs(f(x(:,k))));
   if tol<epsi
         break;
     end
 end
figure(1),subplot(1,2,1), contour(X,Y,Z,fvalue);
hold on, plot(x(1,:), x(2,:), 'o-'), xlabel('x_1'), ylabel('x_2')
subplot(1,2,2), semilogy(1:k+1, fvalue , '-o'), xlabel('k'),...
 ylabel('f(x)')
 %====================================================================
```

3.8 Summary

Necessary and sufficient conditions of optimal solutions in set constrained optimization problems have been presented in this chapter with extensive examples. Considering these conditions, general optimization algorithms using iterative search methods have been introduced. In particular, steepest descent method and Newton's method as two well-known optimization algorithms have been discussed and their implementation has been given. Moreover, the least square optimization that is widely used in engineering applications has been introduced with analytic solution.

3.9 Problems

Problem 1 Consider the function $f(\mathbf{x}) = \mathbf{x}^T \begin{bmatrix} 3 & 1 \\ 1 & 2 \end{bmatrix} \mathbf{x} + \mathbf{x}^T \begin{bmatrix} 3 \\ 1 \end{bmatrix} + 6.$

(a) Find the gradient and Hessian of f at $\mathbf{x} = \begin{bmatrix} 1 \\ -1 \end{bmatrix}$.

(b) Find the directional derivative of f at $\mathbf{x} = \begin{bmatrix} 1 \\ -1 \end{bmatrix}$ with respect to a unit vector in the direction of maximum rate of increase.

(c) Find the point which satisfies the necessary and sufficient conditions of optimality for f. Determine whether this point is a minimum, maximum, or saddle point?

Problem 2 Consider function $f(x_1, x_2) = x_1^2 x_2 + x_2^3 x_1$.

(a) Find the rate of increase of f at point $x = [2, 1]^T$ in the direction $d = [3, 4]^T$.
(b) In what direction does the function f decreases most rapidly at the point $x = [2, 1]^T$?

Problem 3 Find minimum and maximum points of $f(x_1, x_2) = (x_1^2 + x_2^2 - 1)^2 + (x_2^2 - 1)^2$.

Problem 4 Find the minimizer of $f(x_1, x_2, x_3) = (x_1 - 4)^2 + (x_2 - 3)^2 + (x_3 + 4)^2$ using:

(a) The steepest descent method,
(b) The constant step size with $\alpha = 0.1$,

with $\mathbf{x}^0 = \begin{bmatrix} 4 \\ 2 \\ -1 \end{bmatrix}$ and stopping criteria $\|\nabla f(\mathbf{x})\| < 0.001$.

Problem 5 Write a MATLAB program to find the minimizer of Rosenbrock's function:

$$f(\mathbf{x}) = (1 - x_1)^2 + 100(x_2 - x_1^2)^2$$

Use an initial point $\mathbf{x}^0 = \begin{bmatrix} -1 \\ -1 \end{bmatrix}$ and terminate the algorithm when $\|\nabla f(\mathbf{x})\| < 0.001$.

References

1. E.K.P. Chong, S.H. Zak, *An Introduction to Optimization* (Wiley, Hoboken, 2013)
2. A. Beck, *Introduction to Nonlinear Optimization: Theory, Algorithms, and Applications with MATLAB* (SIAM-Society for Industrial and Applied Mathematics, Philadelphia, 2014)
3. J. Nocedal, S. Wright, *Numerical Optimization* (Springer, New York, 2006)

Convex Programming 4

Abstract

One common form of optimization is convex programming that has a wide range of applications in all fields of engineering, in particular electrical engineering. This kind of programming is characterized by some conditions and properties in the construction of functions used in the objective and constraints functions. One major advantage of convex programming is that any local optimal point is also global, which brings forward a great step in the algorithms to solve convex optimization problems. This chapter first defines convex sets and convex functions, using which the definitions of convex and geometric programming problems are determined. To clarify these definitions, appropriate application examples in electrical engineering are given accordingly.

Keywords

Convex programming · Convexity · CVX · Geometric programming · Quasi-convex

4.1 Convex Set

As a basic concept in convex programming, convex set is defined in Definition 1.

Definition 1 A set $\mathcal{C} \subset \mathbb{R}^n$ is convex if

$$\forall \mathbf{x}_1, \mathbf{x}_2 \in \mathcal{C}, 0 \leq \theta \leq 1 \Rightarrow \theta \mathbf{x}_1 + (1-\theta)\mathbf{x}_2 \in \mathcal{C}. \tag{4.1}$$

In other words, $\mathcal{C}$ contains the line segment through any two distinct points in $\mathcal{C}$. A set that is not convex is called a non-convex set.

As an example, given sets on the top of Fig. 4.1 are convex while those in the bottom are non-convex. If we remove the condition $0 \leq \theta \leq 1$ on θ in Definition 1, the set $\mathcal{C}$ is called *affine set*. In other words, the affine set contains the whole line through any two distinct points.

Three well-known convex sets are hyperplanes, halfspaces, and polytopes. These sets are defined in the following and illustrated in Fig. 4.2.

M. Fathi, H. Bevrani, *Optimization in Electrical Engineering*,
https://doi.org/10.1007/978-3-030-05309-3_4

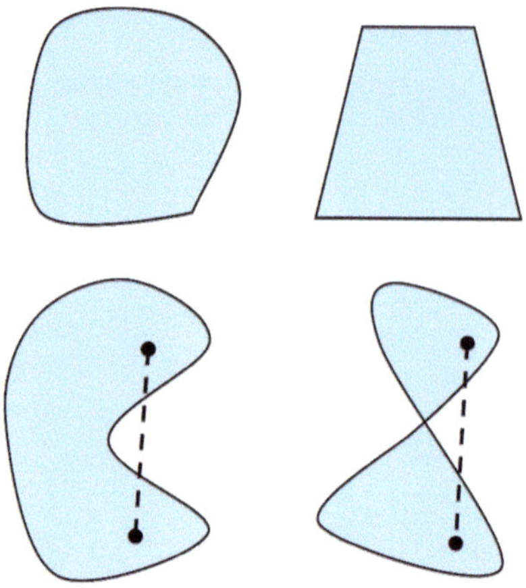

Fig. 4.1 Sets on the top are convex, but sets on the bottom are non-convex

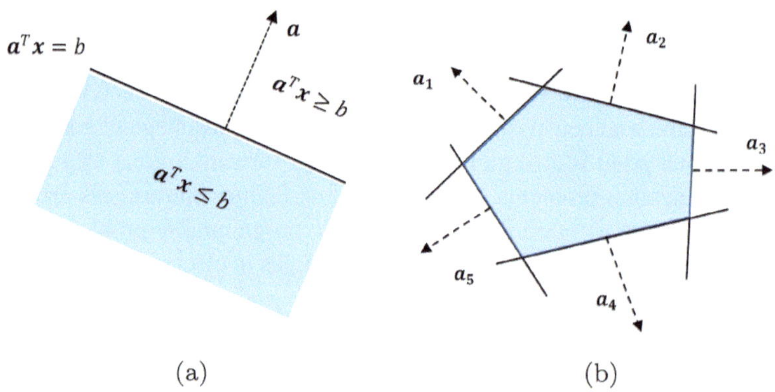

Fig. 4.2 Convex sets: (**a**) hyperplane, halfspace, and (**b**) polytope

- A set of points in the form of $\{\mathbf{x} \in \mathbb{R}^n | \mathbf{a}^T \mathbf{x} = b\}$, where $\mathbf{a} \in \mathbb{R}^n$, $\mathbf{a} \neq 0$, is called a *hyperplane* as in Fig. 4.2a. Vector **a** is said to be the normal vector of the hyperplane. Each hyperplane is a convex set.
- A set of points in the form of $\{\mathbf{x} \in \mathbb{R}^n | \mathbf{a}^T \mathbf{x} \leq b\}$, where $\mathbf{a} \in \mathbb{R}^n$, $\mathbf{a} \neq 0$, is called a *halfspace* as in Fig. 4.2a. Each halfspace is a convex set.
- The intersection of a finite number of hyperplanes and halfspaces is called a polytope as in Fig. 4.2b. Indeed, the polytope is in the form of

$$P = \{\mathbf{x} \in \mathbb{R}^n | \mathbf{a}_i^T \mathbf{x} \leq b_i, i = 1, \ldots, m, \ \mathbf{c}_i^T \mathbf{x} = d_i, i = 1, \ldots, p\}$$

  where $\mathbf{a}_i, \mathbf{c}_i \in \mathbb{R}^n$. Polytopes are convex sets. This is due to the fact that the intersection of a number of convex sets is also convex.

Following the definition of convex set, convex combination is a concept that is used frequently in convex programming. This combination can be used to form a convex set from a finite set of numbers. This is stated in Definition 2.

Definition 2 Convex combination of a set $\mathbf{x}_1, \mathbf{x}_2, \ldots, \mathbf{x}_n$ of points in $\mathbb{R}^n$ is any point in the form of $\theta_1\mathbf{x}_1 + \theta_2\mathbf{x}_2 + \cdots + \theta_n\mathbf{x}_n$, where $\theta_1 + \theta_2 + \cdots + \theta_n = 1$ and $\theta_i \geq 0$. The set of all convex combinations of points in a given set $\mathcal{S}$ is called a convex hull, indicated by conv($\mathcal{S}$) as

$$\text{conv}(\mathcal{S}) = \left\{ \sum_{i=1}^{n} \theta_i \mathbf{x}_i | \mathbf{x}_i \in \mathcal{S}, \sum_{i=1}^{n} \theta_i = 1, \theta_i \geq 0 \right\}. \tag{4.2}$$

Note that the convex hull of a set $\mathcal{S}$ is the smallest convex set that contains $\mathcal{S}$ [1]. Finally, a convex set is closed under convex combinations.

4.2 Convex Function

Having stated the convex set in Definition 1, the next step towards convex programming is the concept of convex function that is stated in Definition 3.

Definition 3 Function $f : \mathbb{R}^n \rightarrow \mathbb{R}$ is convex if D_f (domain of f) is a convex set and

$$\forall \mathbf{x}_1, \mathbf{x}_2 \in D_f, 0 \leq \theta \leq 1 \Rightarrow f(\theta\mathbf{x}_1 + (1-\theta)\mathbf{x}_2) \leq \theta f(\mathbf{x}_1) + (1-\theta) f(\mathbf{x}_2). \tag{4.3}$$

This statement geometrically implies that the line between $(\mathbf{x}_1, f(\mathbf{x}_1))$ and $(\mathbf{x}_2, f(\mathbf{x}_2))$ lies above the graph of f between $\mathbf{x}_1$ and $\mathbf{x}_2$, as shown in Fig. 4.3. Moreover, f is called a *concave* function if $-f$ is convex.

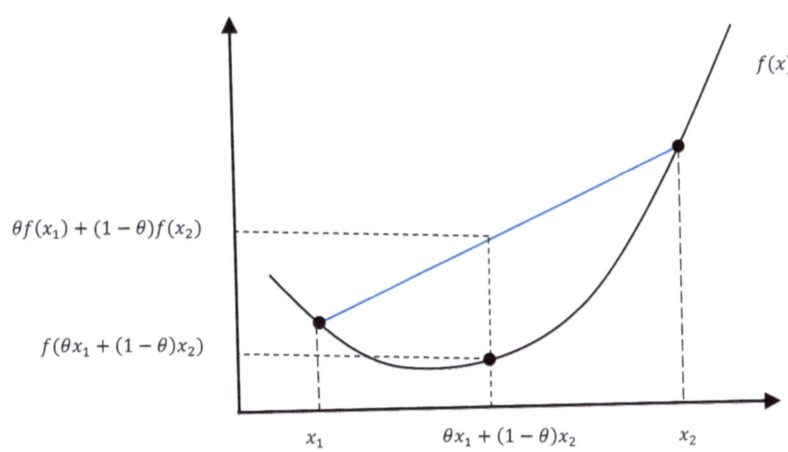

Fig. 4.3 Convex function: the line between any two points lies above the graph

Following are a number of convex function examples:

- Affine function $f(\mathbf{x}) = \mathbf{a}^T\mathbf{x} + b$ for all $\mathbf{a}, \mathbf{x} \in \mathbb{R}^n$ and scalar b.
- Exponential function $f(x) = e^{ax}$ for all $a, x \in \mathbb{R}$.
- Power function $f(x) = x^\alpha$, for $x > 0,\ \alpha \geq 1$ or $\alpha \leq 0$.
- Norm function $\|\mathbf{x}\|_p$ for $p \geq 1$. The proof is based on the triangular inequality in Chap. 2.
- Log-sum of exponential functions $f(\mathbf{x}) = \log(e^{x_1} + e^{x_2} + \cdots + e^{x_n})$.

Similarly, following are a number of concave function examples:

- Affine function $f(\mathbf{x}) = \mathbf{a}^T\mathbf{x} + b$ for all $\mathbf{a}, \mathbf{x} \in \mathbb{R}^n$ and scalar b.
- Power function $f(x) = x^\alpha$, for $x > 0, 0 \leq \alpha \leq 1$.
- Logarithm function $f(x) = \log x$ for $x > 0$.
- Log-determinant function $f(\mathbf{X}) = \log \det(\mathbf{X})$ if $\mathbf{X}$ is a positive semi-definite matrix.

These statements can be verified using the definition of convex function or conditions provided in the upcoming sections.

Based on the provided definition for convex function, first-order and second-order conditions on convex functions are derived and introduced in Sects. 4.2.1 and 4.2.2. The major advantage of these conditions is to check out the convexity of a given function.

4.2.1 First-Order Condition

The first-order condition is based on the first-order derivative of a function and is stated in Theorem 1.

Theorem 1 *Consider a differentiable function f with convex domain D_f. This function is convex if and only if*

$$f(\mathbf{x}) \geq f(\mathbf{x}_0) + \nabla f(\mathbf{x}_0)^T(\mathbf{x} - \mathbf{x}_0) \tag{4.4}$$

holds for all $\mathbf{x}$ and $\mathbf{x}_0 \in D_f$.

Proof First, suppose that f is convex. Therefore, $\forall\, \mathbf{x}, \mathbf{x}_0 \in D_f, 0 \leq \theta \leq 1$, we have $\theta\mathbf{x} + (1-\theta)\mathbf{x}_0 = \mathbf{x}_0 + \theta(\mathbf{x} - \mathbf{x}_0) \in D_f$ and then $f(\mathbf{x}_0 + \theta(\mathbf{x} - \mathbf{x}_0)) \leq (1-\theta)f(\mathbf{x}_0) + \theta f(\mathbf{x})$. Manipulating this derivation and dividing both sides by θ, we derive

$$f(\mathbf{x}) \geq f(\mathbf{x}_0) + \frac{f(\mathbf{x}_0 + \theta(\mathbf{x} - \mathbf{x}_0)) - f(\mathbf{x}_0)}{\theta}. \tag{4.5}$$

Taking the limit as $\theta \to 0$, $\frac{f(\mathbf{x}_0 + \theta(\mathbf{x} - \mathbf{x}_0)) - f(\mathbf{x}_0)}{\theta}$ is the directional derivative of f in direction $\mathbf{x} - \mathbf{x}_0$ that is equal to $\nabla f(\mathbf{x}_0)^T(\mathbf{x} - \mathbf{x}_0)$. Substituting in (4.5), we derive

$$f(\mathbf{x}) \geq f(\mathbf{x}_0) + \nabla f(\mathbf{x}_0)^T(\mathbf{x} - \mathbf{x}_0).$$

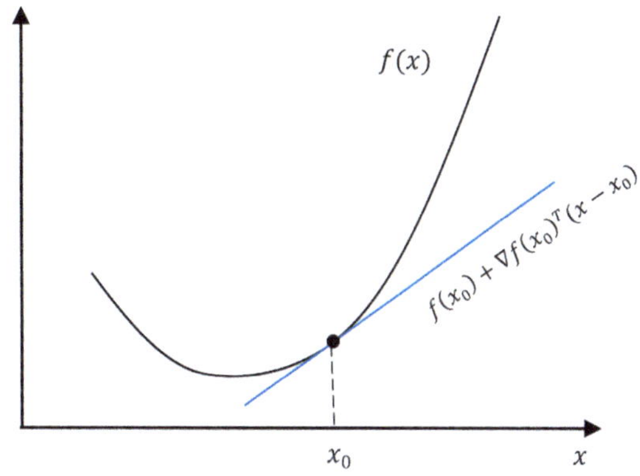

Fig. 4.4 Convex function: first-order approximation is a global under-estimator

This proves the necessary condition. On the other hand, for sufficient condition, let $\mathbf{z} = \theta\mathbf{x}_1 + (1-\theta)\mathbf{x}_2$ for any $\mathbf{x}_1, \mathbf{x}_2 \in D_f$. We can say that

$$f(\mathbf{x}_1) \geq f(\mathbf{z}) + \nabla f(\mathbf{z})^T(\mathbf{x}_1 - \mathbf{z})$$

and

$$f(\mathbf{x}_2) \geq f(\mathbf{z}) + \nabla f(\mathbf{z})^T(\mathbf{x}_2 - \mathbf{z}).$$

Multiplying the first inequality by θ and the second one by $1-\theta$, and adding them together yields

$$f(\mathbf{z}) \leq \theta f(\mathbf{x}_1) + (1-\theta) f(\mathbf{x}_2).$$

This proves the convexity of the function $f(\mathbf{x})$. □

As a more explanation of Theorem 1, the right-hand side of the condition, i.e., $f(\mathbf{x}_0) + \nabla f(\mathbf{x}_0)^T(\mathbf{x} - \mathbf{x}_0)$, is the first-order approximation of f at $\mathbf{x}_0$. Therefore, it is concluded that for a convex function the first-order approximation of f at any point $\mathbf{x}_0$ is a global under-estimator of f at $\mathbf{x}_0$. This is also illustrated in Fig. 4.4. More importantly, if $\nabla f(\mathbf{x}_0) = 0$, we derive $f(\mathbf{x}) \geq f(\mathbf{x}_0)$ for any $\mathbf{x} \in D_f$ and therefore $\mathbf{x}_0$ is the global minimizer.

4.2.2 Second-Order Condition

The second-order condition is based on the second-order derivative of the function and is stated in Theorem 2.

Theorem 2 *A twice-differentiable function f with convex domain D_f is convex if and only if*

$$\nabla^2 f(\mathbf{x}_0) \geq 0 \tag{4.6}$$

for all $\mathbf{x}_0 \in D_f$. In other words, the Hessian matrix is positive semi-definite. Moreover, f is said to be strictly convex if and only if $\nabla^2 f(\mathbf{x}_0) > 0$ for all $\mathbf{x}_0 \in D_f$.

Proof Based on the Taylor series, the second-order approximation of f at $\mathbf{x}_0$ is given by

$$f(\mathbf{x}) \approx f(\mathbf{x}_0) + \nabla f(\mathbf{x}_0)^T(\mathbf{x} - \mathbf{x}_0) + \frac{1}{2}(\mathbf{x} - \mathbf{x}_0)^T \nabla^2 f(\mathbf{x}_0)(\mathbf{x} - \mathbf{x}_0). \tag{4.7}$$

First, suppose that f is convex. Following the first-order condition in Theorem 1, we have $f(\mathbf{x}) \geq f(\mathbf{x}_0) + \nabla f(\mathbf{x}_0)^T(\mathbf{x} - \mathbf{x}_0)$. Considering this inequality along with the second-order approximation in (4.7), it is inferred that $\nabla^2 f(\mathbf{x}_0) \geq 0$.

On the other hand, suppose that $\nabla^2 f(\mathbf{x}_0) \geq 0$. Once again, considering this inequality along with the second-order approximation in (4.7), it is inferred that

$$f(\mathbf{x}) \geq f(\mathbf{x}_0) + \nabla f(\mathbf{x}_0)^T(\mathbf{x} - \mathbf{x}_0)$$

and therefore f is convex as a result of the first-order condition in Theorem 1. □

A similar proof can also be done for the case of strictly convex function by replacing $\geq$ with $>$ in the above statements.

Example 1 Check out the convexity of the following functions:

(a) Quadratic function $f(\mathbf{x}) = (1/2)\mathbf{x}^T\mathbf{P}\mathbf{x} + \mathbf{q}^T\mathbf{x} + r$ ($\mathbf{P}$ is symmetric).
(b) Least square function $f(\mathbf{x}) = \|\mathbf{A}\mathbf{x} - \mathbf{b}\|_2^2$.
(c) Quadratic over linear function $f(x_1, x_2) = x_1^2/x_2$.

Solution

(a) Taking the gradient and Hessian, we have $\nabla f(\mathbf{x}) = \mathbf{P}\mathbf{x} + \mathbf{q}$ and $\nabla^2 f(\mathbf{x}) = \mathbf{P}$. Therefore, f is convex if and only if $\mathbf{P} \geq 0$.
(b) Taking the gradient and Hessian, we derive $\nabla f(x) = 2\mathbf{A}^T(\mathbf{A}\mathbf{x} - \mathbf{b})$ and $\nabla^2 f(\mathbf{x}) = 2\mathbf{A}^T\mathbf{A}$. Since $\mathbf{A}^T\mathbf{A}$ is always a positive semi-definite matrix, f is convex for all $\mathbf{A}$.
(c) The Hessian can be written as

$$\nabla^2 f(x_1, x_2) = \frac{2}{x_2^3}\begin{bmatrix} x_2 \\ -x_1 \end{bmatrix}\begin{bmatrix} x_2 \\ -x_1 \end{bmatrix}^T$$

which is always positive semi-definite for $x_2 > 0$. Therefore, f is convex for all $x_2 > 0$.

□

Example 2 In communication systems, the capacity of an additive white Gaussian noise (AWGN) channel, in bits per second, with transmit power p and bandwidth w is generally written as

$$r(p, w) = \alpha w \log_2(1 + \frac{\beta p}{w}) \tag{4.8}$$

(continued)

where α and β are positive scalars. Determine whether $r(p, w)$ is convex or concave in terms of $\begin{bmatrix} p \\ w \end{bmatrix}$.

Solution Taking the second-order derivative, the Hessian is derived as

$$\nabla^2 r(p, w) = \frac{-\alpha\beta^2}{w \ln(2)\left(1 + \frac{\beta p}{w}\right)^2} \begin{bmatrix} 1 & -\frac{p}{w} \\ -\frac{p}{w} & \frac{p^2}{w^2} \end{bmatrix}$$

which can be written as

$$\nabla^2 r(p, w) = \frac{-\alpha\beta^2}{w \ln(2)\left(1 + \frac{\beta p}{w}\right)^2} \begin{bmatrix} 1 \\ -\frac{p}{w} \end{bmatrix} \begin{bmatrix} 1 \\ -\frac{p}{w} \end{bmatrix}^T.$$

Considering the form of derived $\nabla^2 r(p, w)$, it can be easily verified that it is a negative definite matrix, and therefore $r(p, w)$ is concave in terms of $\begin{bmatrix} p \\ w \end{bmatrix}$. □

4.2.3 Quasi-Convex Function

A real-valued function f defined over a convex set $\mathcal{C}$ is quasi-convex if for all $\mathbf{x}_1, \mathbf{x}_2 \in \mathcal{C}$ and $0 \leq \theta \leq 1$, we have

$$f(\theta\mathbf{x}_1 + (1-\theta)\mathbf{x}_2) \leq \max\{f(\mathbf{x}_1), f(\mathbf{x}_2)\}. \tag{4.9}$$

In other words, a point between any two other points does not give a higher function value than both of the other points. An alternative way of defining a quasi-convex function is that each *α-sublevel set* of f denoted by

$$\mathcal{S}_\alpha^- = \{\mathbf{x} \in \mathcal{C} | f(\mathbf{x}) \leq \alpha\} \tag{4.10}$$

is a convex set.

Quasi-convexity is an extension of convexity. In other words, all convex functions are quasi-convex, however, not all quasi-convex functions are convex. Moreover, a function f is said to be *quasi-concave* if $-f$ is quasi-convex. In other words, for a quasi-concave function, *α-superlevel sets* indicated by $\mathcal{S}_\alpha^+ = \{\mathbf{x} \in \mathcal{C} | f(\mathbf{x}) \geq \alpha\}$ are convex sets.

As an example, consider the function in Fig. 4.5. This function is not a convex function, instead it is a quasi-convex function. Moreover, the function in Fig. 4.6 is neither convex nor quasi-convex. For example, the set of points in which $f(\mathbf{x}) \leq 0.5$ is the union of two bolded intervals, which is not a convex set.

Example 3 Figure 4.7 illustrates the contour plots of $f(\mathbf{x}) = x_2 \log\left(1 + \frac{x_1}{x_2}\right)$, where $D_f = \mathbb{R}_+^2$. Using this figure, determine the convexity of $f(\mathbf{x})$.

Solution As seen, the superlevel sets of this function, i.e., $\{\mathbf{x} \in \mathbb{R}_+^2 | f(\mathbf{x}) \geq \alpha\}$, are convex set. Therefore, the function is quasi-concave. In order to check out if the function is concave or not,

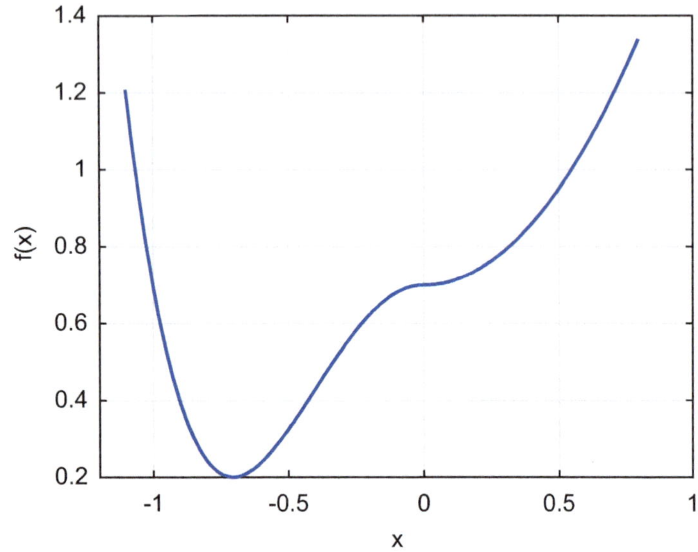

Fig. 4.5 Quasi-convex but not convex function

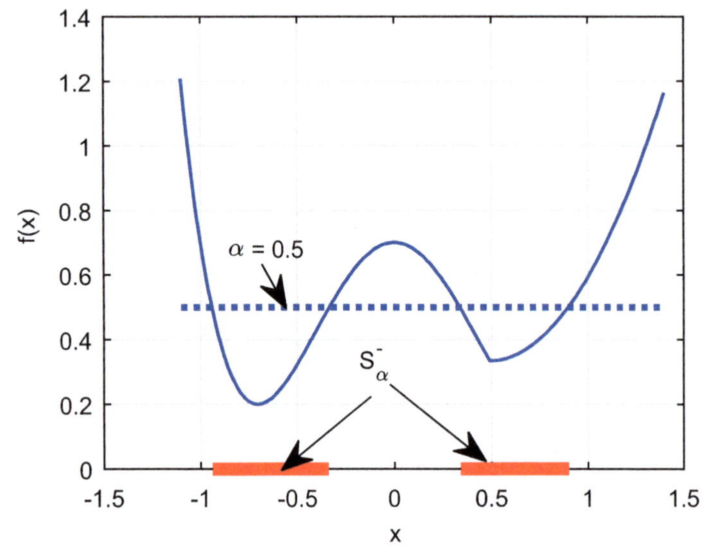

Fig. 4.6 Neither quasi-convex nor quasi-concave function

consider to pass any line through the level sets and indicate the intersections as $(\mathbf{x}_i, f(\mathbf{x}_i))$. Now, consider to plot $f(\mathbf{x}_i)$'s versus $\mathbf{x}_i$'s. It is observed that a unit increase in the function value occurs by greater steps of increase in $\mathbf{x}$, which results in a downward curvature. This is the property of concave functions. Therefore, the function is concave as well. □

Note that the concavity of the channel capacity function $r(p, w)$ in Example 2 can also be verified using the concavity of $f(\mathbf{x}) = x_2 \log\left(1 + \frac{x_1}{x_2}\right)$ in Example 3 by some manipulations.

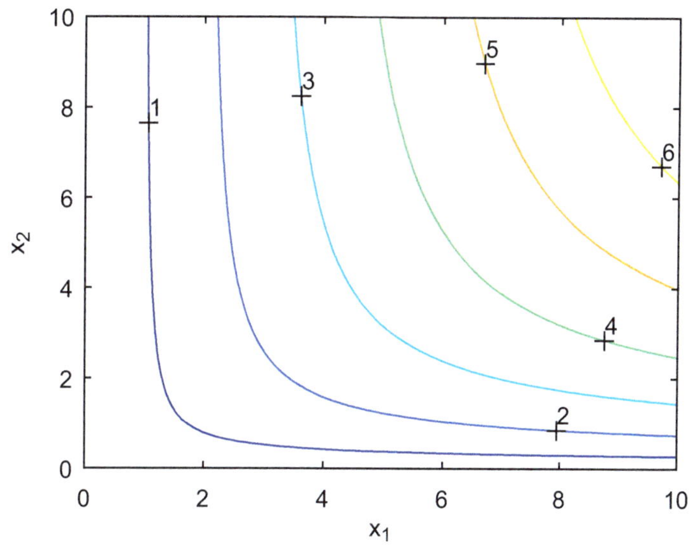

Fig. 4.7 Concave function: $f(\mathbf{x}) = x_2 \log\left(1 + \frac{x_1}{x_2}\right)$

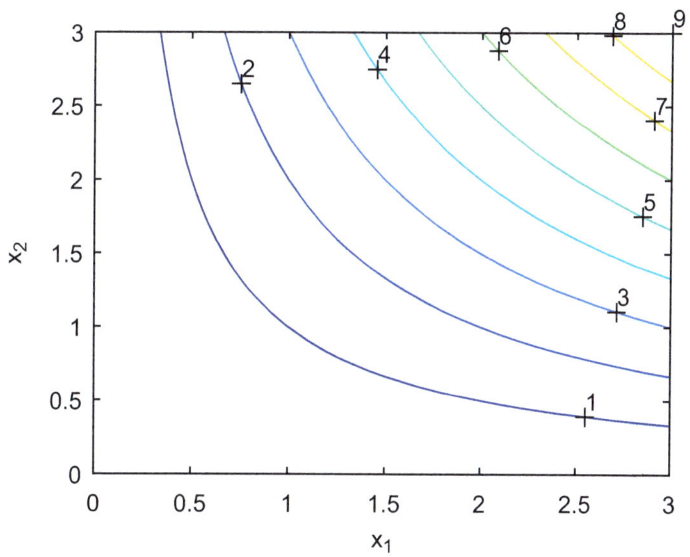

Fig. 4.8 Quasi-concave function: $f(\mathbf{x}) = x_1 x_2$

Example 4 Check out the convexity of $f(\mathbf{x}) = x_1 x_2$, where $D_f = \mathbb{R}_+^2$.

Solution The contour plots of this function are shown in Fig. 4.8. As shown, the superlevel sets, i.e., $\{\mathbf{x} \in \mathbb{R}_+^2 | x_1 x_2 \geq \alpha\}$, are convex. Therefore, the function is quasi-concave. As in the solution of Example 3, if we pass any line through the level sets and plot the function values versus $\mathbf{x}$'s, we observe that a unit increase in the function value occurs by smaller steps of increase in $\mathbf{x}$, which results in an upward curvature. This property is not seen in concave functions. Therefore, the function is quasi-convex but not concave. □

4.2.4 Convexity Preservation

In general, checking out the convexity of a function is not trivial. Therefore, it is advantageous to investigate the mathematical operations that preserve the convexity property. Following are a number of these operations. Verifying these operations using convexity definition is left as an exercise to the reader.

- *Non-negative weighted sum:* If $f_1, f_2, \ldots, f_n$ are convex functions and $w_1 \geq 0, \ldots, w_n \geq 0$, then

$$w_1 f_1 + w_2 f_2 + \cdots + w_n f_n \tag{4.11}$$

  is a convex function. Similarly, a non-negative weighted sum of concave functions is also concave.
- *Composition with an affine function:* Suppose $f : \mathbb{R}^n \to \mathbb{R}$, $\mathbf{A} \in \mathbb{R}^{m\times n}$ and $\mathbf{b} \in \mathbb{R}^m$. If f is convex, then $f(\mathbf{Ax}+\mathbf{b})$ is also convex. For example, $f(\mathbf{x}) = \|\mathbf{Ax}+\mathbf{b}\|$ is convex as the norm functions are convex.

Example 5 Check out the convexity of $f(\mathbf{x}) = 5e^{6x_1-3x_2+4} + 2(3x_1+x_2-3)^2$.

Solution The convex function e^z is composed with the affine function $6x_1-3x_2+4$, and therefore $e^{6x_1-3x_2+4}$ is convex. Similarly, the convex function z^2 is composed with the affine function $3x_1+x_2-3$ and so it is convex. Moreover, non-negative weighted sum of convex functions is convex and therefore $f(\mathbf{x})$ is also convex. □

- *Pointwise maximum:* If $f_1, f_2, \ldots, f_n$ are convex functions, then

$$f(\mathbf{x}) = \max\{f_1(\mathbf{x}), \ldots, f_n(\mathbf{x})\} \tag{4.12}$$

  is also convex. Conversely, the pointwise minimum of a number of concave functions is concave.
- *Pointwise supremum:* This is an extension of pointwise maximum operation. If $f(\mathbf{x},\mathbf{y})$ is convex in $\mathbf{x}$ for each $\mathbf{y}$ in a given set, then

$$g(\mathbf{x}) = \sup_{\mathbf{y}} f(\mathbf{x},\mathbf{y}) = \max\{f(\mathbf{x},\mathbf{y}_1), f(\mathbf{x},\mathbf{y}_2), \ldots\} \tag{4.13}$$

  is convex.

4.3 Standard Optimization Problem

Having introduced convex sets and convex functions, it is turn to discuss the structure and properties of convex problems. Prior to this, following the set constrained problem in Chap. 3, the standard optimization problem is considered in the form of

$$\min_{\mathbf{x}} \; f_0(\mathbf{x}) \tag{4.14a}$$

$$\text{subject to} \;\; f_i(\mathbf{x}) \leq 0, \quad i = 1, \ldots, m \tag{4.14b}$$

$$h_i(\mathbf{x}) = 0, \quad i = 1, \ldots, p \tag{4.14c}$$

in which

- $\mathbf{x} \in \mathbb{R}^n$ is the vector of optimization variables.
- $f_0 : \mathbb{R}^n \to \mathbb{R}$ is the objective function.
- $f_i : \mathbb{R}^n \to \mathbb{R}, i = 1, \ldots, m$ are inequality constraint functions.
- $h_i : \mathbb{R}^n \to \mathbb{R}, i = 1, \ldots, p$ are equality constraint functions.

The problem domain is the set of points for which the objective and constraint functions are defined. This set is indicated by

$$D = \cap_{i=0}^{m} D_{f_i} \cap_{i=1}^{p} D_{h_i} \tag{4.15}$$

where D_{f_i} and D_{h_i} are domains of objective, inequality and equality constraint functions, respectively. Any $\mathbf{x} \in D$ that satisfies the constraints (4.14b) and (4.14c) is called a feasible point. The set of all feasible points is called *feasible region* and is denoted by Ω. Furthermore, the *optimal value* of this problem is defined as

$$p^* = \inf\{f_0(\mathbf{x})|\mathbf{x} \in \Omega\} \tag{4.16}$$

and a feasible point $\mathbf{x}^*$ is *optimal* if $f_0(\mathbf{x}^*) = p^*$. If $p^* = \infty$, the problem is said to be infeasible, i.e., $\Omega = \emptyset$. Moreover, if $p^* = -\infty$, the problem is said to be unbounded below.

Solving a standard optimization problem in general is difficult. However, specific categories of optimization problems and in particular convex problems can be solved efficiently due to special form of objective and constraint functions [1, 2]. The structure of convex problems and corresponding optimality conditions are investigated in Sect. 4.4.

4.4 Convex Problem

Following the statement of the standard optimization problem in (4.14), the convex optimization problem is stated in the form of

$$\min_{\mathbf{x}} \; f_0(\mathbf{x}) \tag{4.17a}$$

$$\text{subject to } \; f_i(\mathbf{x}) \leq 0, \quad i = 1, \ldots, m \tag{4.17b}$$

$$\mathbf{a}_i^T \mathbf{x} = b_i, \quad i = 1, \ldots, p \tag{4.17c}$$

in which $f_0, f_1, \ldots, f_m$ are *convex* functions and equality constraint functions $\mathbf{a}_i^T \mathbf{x} - b_i$ are *affine*. Indeed, a convex problem is a special form of a standard optimization problem in which the objective and inequality constraint functions are convex, and moreover equality constraint functions are affine. These characteristics result in a convex feasible region for convex problems, as stated in Theorem 3.

Theorem 3 *The feasible region of the convex optimization problem is a convex set.*

Proof The feasible region Ω of a convex problem is the intersection of problem domain D in (4.15) and the set of points satisfying inequality and equality constraints. This set is in turn the intersection of convex sublevel sets $\{\mathbf{x}|f_i(\mathbf{x}) \leq 0\}$ for $i = 1, \ldots, m$ and convex hyperplanes $\{\mathbf{x}|\mathbf{a}_i^T \mathbf{x} = b_i\}$ for

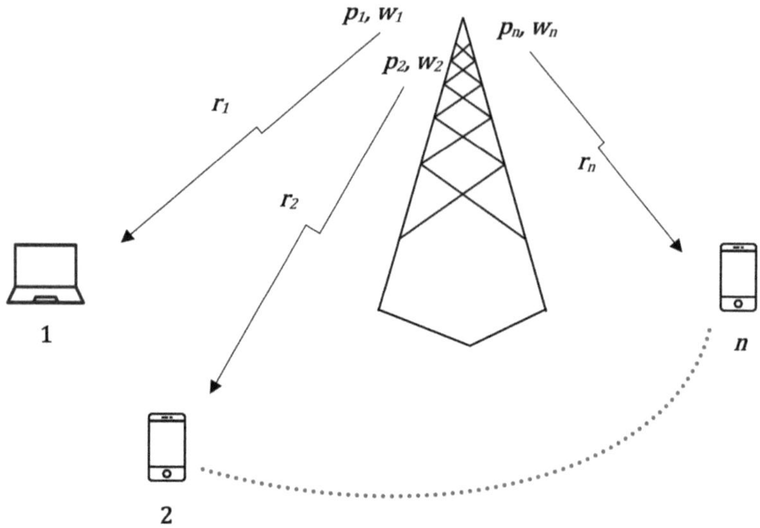

Fig. 4.9 Power and bandwidth allocation in wireless networks

$i = 1, \ldots, p$. We know that the intersection of a number of convex sets is certainly a convex set. Consequently, Ω is convex. □

Example 6 Consider the wireless network in Fig. 4.9 consisting of n users. The power and bandwidth resources in the network are to be allocated to the users. Allocating power p_i and bandwidth w_i to user i, the achieved transmit rate is $r_i = \alpha_i w_i \log_2\left(1 + \frac{\beta_i p_i}{w_i}\right)$, where α_i and β_i are positive scalars. Let W be the available bandwidth in the network.

Now consider the problem of allocating power and bandwidth jointly in the network such that the transmit rate of each user i must not be less than r_i^{th} and an economic measure of transmit powers to be minimized. Formulate this problem as a convex programming.

Solution This problem can be formulated as in (4.18a) with a squared cost function of powers.

$$\min_{\mathbf{p},\mathbf{w}} \left(\sum_{i=1}^{n} p_i\right)^2 \tag{4.18a}$$

$$\text{subject to } r_i^{\text{th}} - \alpha_i w_i \log_2\left(1 + \frac{\beta_i p_i}{w_i}\right) \leq 0, \qquad i = 1, \ldots, n \tag{4.18b}$$

$$w_1 + w_2 + \cdots + w_n = W, \tag{4.18c}$$

$$-p_i \leq 0, \qquad i = 1, \ldots, n \tag{4.18d}$$

$$-w_i \leq 0. \qquad i = 1, \ldots, n \tag{4.18e}$$

The objective (4.18a) is a quadratic and positive function thus it is convex. Considering the concavity of the channel capacity function in Example 2, it is inferred that left-hand side of constraint (4.18b) is

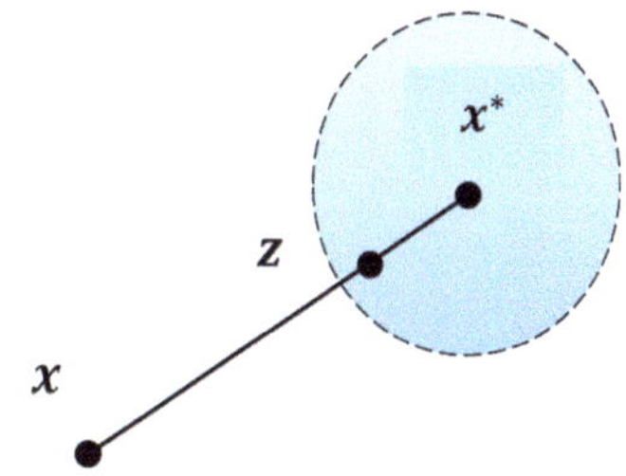

Fig. 4.10 $\mathbf{x}^*$ as a local and global optimal point in the feasible set of a convex problem

a convex function. Moreover, the equality constraint (4.18c) is affine. Finally, the last two constraints are also convex. Consequently, (4.18) is a convex problem. □

Theorems 4 and 5 discuss optimality properties and conditions for convex problems, which are frequently employed by standard algorithms designed of convex programming [2–4].

Theorem 4 *Any local optimal point of a convex problem is also globally optimal.*

Proof Assume that $\mathbf{x}^*$ be a local but not global optimal point. As it is not globally optimal, there is a point $\mathbf{x}$ such that $f(\mathbf{x}) < f(\mathbf{x}^*)$. On the other hand, due to the local optimality of $\mathbf{x}^*$, $f(\mathbf{x}^*)$ has the smallest value in a neighborhood of $\mathbf{x}^*$ as in Fig. 4.10. Now, let $\mathbf{z} = \theta\mathbf{x}^* + (1-\theta)\mathbf{x}$ be a close enough point to $\mathbf{x}^*$ such that it is located in this neighborhood. As a result of convexity, we have $f(\mathbf{z}) \leq \theta f(\mathbf{x}^*) + (1-\theta)f(\mathbf{x})$, which concludes in $f(\mathbf{z}) < f(\mathbf{x}^*)$ due to $f(\mathbf{x}) < f(\mathbf{x}^*)$. This conclusion contradicts the local optimality of $\mathbf{x}^*$. Therefore, $\mathbf{x}^*$ is also a global optimal point. □

Recall that we discussed the optimality conditions of set constrained optimization problems in Chap. 3. Here, in Theorem 5, we revise the first-order necessary condition for convex problems. As shown, unlike general optimization problems, this condition becomes both necessary and sufficient condition for optimality in convex problems.

Theorem 5 *Consider a convex problem with objective function f_0 and feasible set Ω. Then, $\mathbf{x}^* \in \Omega$ is optimal if and only if $\nabla f_0(\mathbf{x}^*)^T(\mathbf{x}-\mathbf{x}^*) \geq 0$ for all $\mathbf{x} \in \Omega$.*

Proof First, suppose that $\nabla f_0(\mathbf{x}^*)^T(\mathbf{x}-\mathbf{x}^*) \geq 0$ for all $\mathbf{x} \in \Omega$. Due to the convexity of f_0, we also have $f_0(\mathbf{x}) \geq f_0(\mathbf{x}^*) + \nabla f_0(\mathbf{x}^*)^T(\mathbf{x}-\mathbf{x}^*)$. From these statements, we conclude that $f_0(\mathbf{x}) \geq f_0(\mathbf{x}^*)$ for all $\mathbf{x} \in \Omega$. Therefore, $\mathbf{x}^*$ is an optimal global minimizer. Conversely, let $\mathbf{x}^*$ be an optimal point. Following the first-order necessary condition in Chap. 3, the directional derivative of f_0 is non-negative at $\mathbf{x}^*$, i.e., $\nabla f_0(\mathbf{x}^*)^T(\mathbf{x}-\mathbf{x}^*) \geq 0$ for all $\mathbf{x} \in \Omega$. □

It is of high importance that the mentioned optimality condition in Theorem 5 satisfies both necessary and sufficient condition for convex problems. As a special case, for an unconstrained convex problem, i.e., $\Omega = \mathbb{R}^n$, $\mathbf{x}^*$ is optimal if and only if $\nabla f_0(\mathbf{x}^*) = 0$. For example, the minimizer of the quadratic function $f_0(\mathbf{x}) = (1/2)\mathbf{x}^T\mathbf{P}\mathbf{x} + \mathbf{q}^T\mathbf{x} + r$ ($\mathbf{P} > 0$ and symmetric) is derived using $\nabla f(\mathbf{x}^*) = \mathbf{P}\mathbf{x}^* + \mathbf{q} = 0$, i.e., $\mathbf{x}^* = -\mathbf{P}^{-1}\mathbf{q}$.

Having the concept of convex programming, in the following subsections, a number of optimization programming categories which fall into the area of convex programming are introduced. It is noteworthy that the discussion on solving optimization problems is left to Chap. 5.

4.4.1 Linear Programming

An optimization problem with affine objective and constraint functions in the form of

$$\min_{\mathbf{x}} \ \mathbf{c}^T\mathbf{x} \tag{4.19a}$$

$$\text{subject to } \ \mathbf{g}_i^T\mathbf{x} \leq d_i, \quad i = 1, \ldots, m \tag{4.19b}$$

$$\mathbf{a}_i^T\mathbf{x} = b_i, \quad i = 1, \ldots, p \tag{4.19c}$$

is called a *linear program*, where $\mathbf{c}$, $\mathbf{g_i}$ and $\mathbf{a}_i \in \mathbb{R}^n$. Since this problem satisfies the requirements of convex problem, all linear optimization problems are convex and therefore can be solved using convex programming algorithms.

As an important point, the feasible region of a linear problem is a polytope determined by affine constraints (4.19b) and (4.19c). It can be shown that the optimal solution always lies on a vertex of the polytope. Based on this fact, numerical solutions of linear problems have extensively been discussed in the literature [5].

4.4.2 Quasi-Convex Problem

Following our discussion on quasi-convex functions in Sect. 4.2.3, the problem in the form of

$$\min_{\mathbf{x}} \ f_0(\mathbf{x}) \tag{4.20a}$$

$$\text{subject to } \ f_i(\mathbf{x}) \leq 0, \quad i = 1, \ldots, m \tag{4.20b}$$

$$\mathbf{a}_i^T\mathbf{x} = b_i, \quad i = 1, \ldots, p \tag{4.20c}$$

is quasi-convex if f_0 is quasi-convex and $f_1, \ldots, f_m$ are convex.

Indeed, this is a generalization on convex problems, in which the objective function is not required to be convex. However, the other requirements are the same as in convex problems. Furthermore, unlike the convex problem, quasi-convex problem can have locally optimal points that are not globally optimal. This is evident from Fig. 4.5.

Theorem 6 presents the optimality condition for quasi-convex problems.

Theorem 6 *Consider a quasi-convex problem with objective function f_0 and feasible set Ω. If $\nabla f_0(\mathbf{x}^*)^T(\mathbf{x} - \mathbf{x}^*) \geq 0$ for all $\mathbf{x} \in \Omega$, then $\mathbf{x}^* \in \Omega$ is optimal.*

This condition is different from optimality condition for convex problems in Theorem 5. First, unlike the optimality condition for convex problems, the optimality condition for quasi-convex problems is only a sufficient condition. In other words, there might be an optimal $\mathbf{x}^*$ for which optimality condition does not hold. Second, this optimality condition requires nonzero $\nabla f_0(\mathbf{x}^*)$.

4.4.3 Geometric Programming

Geometric programming is a kind of optimization problem with a special form of objective and constraint functions that is widely used in engineering applications [6]. This problem is not convex in general. However, it can be converted into a convex problem by a change of variables and a transformation of objective and constraint functions. The solutions follow the same algorithms of convex problems. Even large-scale geometric problems can be solved efficiently using numerical methods [7].

Prior to introducing a geometric problem and the way of transforming it, let us consider Definitions 4 and 5.

Definition 4 A real-valued function $f : \mathbb{R}^n_+ \rightarrow \mathbb{R}$ with non-negative domain and in the form of

$$f(\mathbf{x}) = cx_1^{a_1}x_2^{a_2}\dots x_n^{a_n} \tag{4.21}$$

is called a *monomial* if $c > 0$ and every exponent a_i be a real number.

For example, x_1x_2, $\frac{x_1}{x_2}$, $0.6x_1^2x_2^{0.5}$ are monomials in $\mathbf{x}$ and $-3x_1x_2$ is not monomial.

Definition 5 Sum of a number of monomial functions in the form of

$$f(\mathbf{x}) = \sum_{k=1}^{K} c_k x_1^{a_{1k}} x_2^{a_{2k}} \dots x_n^{a_{nk}} \tag{4.22}$$

is called a *posynomial* function.

For example, $0.5x_1x_2^{-1} + x_2^{0.2}$ is a posynomial in $\mathbf{x}$ and $x_1 - x_2$ is not posynomial. Posynomials are closed under addition and positive multiplication but not division.

Following Definitions 4 and 5, a problem in the form of

$$\min_{\mathbf{x}} \ f_0(\mathbf{x}) \tag{4.23a}$$

$$\text{subject to} \ f_i(\mathbf{x}) \leq 1, \quad i = 1, \dots, m \tag{4.23b}$$

$$h_i(\mathbf{x}) = 1, \quad i = 1, \dots, p \tag{4.23c}$$

is called a *geometric problem* if f_i's are posynomial and h_i's are monomial functions. As an example, following is a geometric problem:

$$\min_{\mathbf{x}} \ 0.5x_1x_2x_3^{-1} + 3x_1^2x_2 + 6x_1^{-1}x_3 \tag{4.24a}$$

$$\text{subject to } x_1^2 + x_2x_3 \le 1, \tag{4.24b}$$

$$2x_1x_2x_3 = 1. \tag{4.24c}$$

As a special case, we can maximize a nonzero monomial objective function by minimizing its inverse that is also a monomial.

Since posynomials are not convex functions, a geometric programming is not a convex problem in general. The main approach to solve a geometric problem is to convert it into a convex optimization problem with convex objective and inequality functions, and affine equality constraints. This conversion is based on a logarithmic change of variables, and taking a logarithmic transformation of objective and constraint functions. In particular, this is done by changing each variable x_i with $y_i \triangleq \log x_i$. The resulting optimization variable is denoted by $\mathbf{y} \triangleq [y_1, \ldots, y_n]^T$.

To turn the problem into a convex form, instead of minimizing the posynomial function $f_0(\mathbf{x})$, its logarithm $\log f_0(e^{\mathbf{y}})$ is minimized. The notation $e^{\mathbf{y}}$ means component-wise exponentiation, i.e., $(e^{\mathbf{y}})_i = e^{y_i}$. Moreover, inequality constraints $f_i(\mathbf{x}) \le 1$ are replaced with equivalent ones $\log f_i(e^{\mathbf{y}}) \le 0$.

Having deployed the aforementioned transformations, the posynomial $f_i(\mathbf{x}) = \sum_{k=1}^{K} c_k x_1^{a_{1k}} x_2^{a_{2k}} \ldots x_n^{a_{nk}}$ is transformed to

$$\log f_i(e^{\mathbf{y}}) = \log\left(\sum_{k=1}^{K} e^{\mathbf{a}_{ik}^T\mathbf{y}+b_{ik}}\right) \tag{4.25}$$

where $\mathbf{a}_{ik}^T \triangleq [a_{1k}, \ldots, a_{nk}]$ and $b_{ik} \triangleq \log c_k$. The function in (4.25) is a log-sum-exp function that is a convex function. Similarly, equality constraints $h_i(\mathbf{x}) = 1$ are replaced with $\log h_i(e^{\mathbf{y}}) = 0$. In other words, $h_i(\mathbf{x}) = c_i x_1^{a_{1i}} x_2^{a_{2i}} \ldots x_n^{a_{ni}}$ is transformed to

$$\log h_i(e^{\mathbf{y}}) = \mathbf{a}_i^T\mathbf{y} + b_i \tag{4.26}$$

where $\mathbf{a}_i^T \triangleq [a_{1i}, \ldots, a_{ni}]$ and $b_i \triangleq \log c_i$. This is an affine function. Performing the operations in (4.25) and (4.26) on problem (4.23) results in a convex problem in the form of

$$\min_{\mathbf{y}} \ \log\left(\sum_{k=1}^{K} e^{\mathbf{a}_{0k}^T\mathbf{y}+b_{0k}}\right) \tag{4.27a}$$

$$\text{subject to } \log\left(\sum_{k=1}^{K} e^{\mathbf{a}_{ik}^T\mathbf{y}+b_{ik}}\right) \le 0, \quad i=1,\ldots,m \tag{4.27b}$$

$$\mathbf{a}_i^T\mathbf{y} + b_i = 0, \quad i = 1, \ldots, p \tag{4.27c}$$

with new optimization vector $\mathbf{y}$. Unlike the original geometric problem in (4.23), problem (4.27) is convex and can be solved using convex optimization algorithms [6]. As an example, following is the transformation of problem (4.24) into the convex form:

$$\min_{\mathbf{y}} \ \log\left(e^{y_1+y_2-y_3+\ln(0.5)} + e^{2y_1+y_2+\ln(3)} + e^{-y_1+y_3+\ln(6)}\right) \tag{4.28a}$$

$$\text{subject to } \log\left(e^{2y_1} + e^{y_2+y_3}\right) \le 0, \tag{4.28b}$$

$$y_1 + y_2 + y_3 + \ln(2) = 0, \tag{4.28c}$$

which is a convex function.

4.5 Application Examples

So far, the fundamentals of convex programming and the way of recognizing a convex problem have been discussed. For now, to highlight the importance of this programming and its application in the field of electrical engineering [8, 9], several application examples are given in this section.

The approaches towards analytic solution of convex programming are left to Chap. 5. Alternatively, the solution of the given examples is presented using the state-of-the-art solvers deployed by software packages [10–12]. In this book, CVX [10] as a well-known tool to solve convex programming is adopted. CVX is a MATLAB-based modeling system for convex optimization allowing objective and constraints to be specified using MATLAB expression syntax. In CVX, functions and sets are built up from a small set of rules from convex analysis. This tool also supports geometric programming. For more information, see http://cvxr.com/cvx/ and included user guides.

4.5.1 Wireless Communication Channel

Here, to get started with CVX, finding the parameters of the wireless communication channel stated in Chap. 3, which is an unconstrained convex problem, is given in the following:

```
%==============================================
A=[1 -10*log10(10); 1 -10*log10(20); 1 -10*log10(50);...
1 -10*log10(100); 1 -10*log10(300) ];
b=[-70; -75; -90; -110; -125];

cvx_begin
variable x(2)
minimize norm(A*x-b)
cvx_end
```

```
Status: Solved
Optimal value (cvx_optval): +7.52421

x =

  -26.7440
    3.9669

%==============================================
```

The provided result is the same as in Chap. 3.

4.5.2 Joint Power and Bandwidth Allocation

The problem of joint power and bandwidth allocation, raised in Example 6, is also solved using CVX for $n = 2$ users and assumed values for W, r^{th}, α, and β in the following. Just note that `rel_entr(x,y)` in CVX is equivalent to $x \log\left(\frac{x}{y}\right)$.

```
%=========================================
W = 1;
rth = [2; 1];
a= [1; 1];
b = [1; 0.5];
cvx_begin
variables p(2) w(2)
minimize sum(p)^2
subject to
for i=1:2
rth(i) - (-a(i)/log(2))*rel_entr(w(i), w(i)+b(i)*p(i)) <= 0
end
sum(w) == W
w >= 0
p >= 0
cvx_end
p
w
```

```
Status: Solved
Optimal value (cvx_optval): +83.3532

p =

    5.1567
    3.9731


w =

    0.6219
    0.3781
%=========================================
```

4.5.3 Power Control in Wireless Communication

Power control has always been a challenging issue in wireless communication as the power budget is limited in wireless transmitting nodes [13, 14]. Moreover, the transmit power by an individual node causes interference to the other nodes and thus degrades the quality of their communication links. A power control problem usually arising in multi-point to multi-point communication networks is investigated here.

Consider a wireless communication network consisting of a set $\mathcal{N}_T$ of m transmitting nodes and a set $\mathcal{N}_R$ of m receiving nodes. Each node $i \in \mathcal{N}_T$ is going to establish a connection and transmit data to the same node index $i \in \mathcal{N}_R$. The channel gain between any pair of transmitting node $i \in \mathcal{N}_T$ and receiving node $j \in \mathcal{N}_R$ is denoted by $\mathbf{H}_{ij}$, which is a function of path loss, shadowing, and fading.

Under the assumption of transmit power p_i from node $i \in \mathcal{N}_T$, the signal strength at its corresponding receiving node $i \in \mathcal{N}_R$ is $\mathbf{H}_{ii} p_i$ and the interference plus noise is $\sum_{j=1, j\neq i}^{m} \mathbf{H}_{ji} p_j + \sigma^2$, where $\sum_{j=1, j\neq i}^{m} \mathbf{H}_{ji} p_j$ is the interference from simultaneous transmissions by the other nodes in $\mathcal{N}_T$ and σ^2 is the noise power. The received service by each node i is a function of its signal to interference plus noise ratio (SINR) determined by

$$\gamma_i = \frac{\mathbf{H}_{ii} p_i}{\sum_{j=1, j\neq i}^{m} \mathbf{H}_{ji} p_j + \sigma^2}. \tag{4.29}$$

One desired objective in the network is to minimize the total transmit power by the nodes while satisfying a given SINR for each node i, i.e., $\gamma_i \geq \gamma_i^{\text{th}}$. The problem of determining transmit powers $\mathbf{p} = \{p_i\}_{i=1}^{m}$ can be formulated as

$$\min_{\mathbf{p}} \ p_1 + p_2 + \cdots + p_m \tag{4.30a}$$

subject to

$$\frac{\mathbf{H}_{ii} p_i}{\sum_{j=1, j\neq i}^{m} \mathbf{H}_{ji} p_j + \sigma^2} \geq \gamma_i^{\text{th}}, \qquad i = 1, \ldots, m \tag{4.30b}$$

$$p_i \geq 0. \qquad i = 1, \ldots, m. \tag{4.30c}$$

The objective of this power control problem is an affine function. Moreover, the constraint in (4.30b) can be rewritten as

$$\sum_{j=1, j\neq i}^{m} \gamma_i^{\text{th}} \mathbf{H}_{ji} p_j - \mathbf{H}_{ii} p_i + \gamma_i^{\text{th}} \sigma^2 \leq 0$$

which is also an affine function. Therefore, the problem in (4.30) is linear programming and so a convex problem. This problem can be solved using CVX package in MATLAB with typical given channel gains $\mathbf{H}_{ij} = e^{-|i-j|}$ and given $\gamma^{\text{th}} = [15, 12, 9, 6]$.

```
%=============================================
m = 4;
H = [];
SINR = [15; 12; 9; 6];
noise_pow =  1e-9;
for i=1:m
    for j=1:m
        H(i,j) = exp(-abs(i-j));
    end
end
d = diag(H);
cvx_begin
variable p(m,1)
minimize sum(p)
subject to
(H-diag(d))'*p.*SNR - d.*p + noise_pow*SINR <= 0
```

```
p >= 0;
cvx_end
p
```

```
Status: Inaccurate/Solved
Optimal value (cvx_optval): +3.98981e-09
 p =

   1.0e-08 *

    0.1408
    0.0668
    0.0582
    0.1332
%===========================================
```

As another problem, in the mentioned power control scenario, assume that the objective is to maximize the minimum SINR over the network receivers. In other words, the objective is to

$$\max_{\mathbf{p}} \quad \min_{i} \quad \gamma_i \tag{4.31}$$

such that $p_i \leq p_{\max}$ for all i. In other words, this objective aims to offer an equal service to all the receiving nodes. To formulate this objective into an optimization problem, let $t \triangleq \min_{i} \gamma_i$ be the minimum SINR and thus $\gamma_i \geq t$ for all i. Therefore, the problem can be formulated as

$$\max_{\mathbf{p}} \ t \tag{4.32a}$$

$$\text{subject to} \quad \frac{\mathbf{H}_{ii} p_i}{\sum_{j=1, j\neq i}^{m} \mathbf{H}_{ji} p_j + \sigma^2} \geq t, \qquad i = 1, \ldots, m \tag{4.32b}$$

$$0 \leq p_i \leq p_{\max}, \qquad i = 1, \ldots, m. \tag{4.32c}$$

Note that the objective in (4.32a) can be written as $\min_{\mathbf{p}} \frac{1}{t}$ that is a monomial. Moreover, the constraint in (4.32b) can be rewritten as

$$\frac{t\left(\sum_{j=1, j\neq i}^{m} \mathbf{H}_{ji} p_j + \sigma^2\right)}{\mathbf{H}_{ii} p_i} \leq 1$$

where the left-hand side derivation of this constraint is a posynomial. Considering the monomial functions in the objective (4.32a) and constraint (4.32c), it implies that this power control problem is a geometric programming. This can be solved in MATLAB using the gp switch of the CVX package. Following is a numerical solution for this problem assuming $p_{\max} = 1$:

```
%===========================================
m = 4;
H = [];
noise_pow = 1e-9;
```

```
pmax = 1;
for i=1:m
    for j=1:m
        H(i,j) = exp(-abs(i-j));
    end
end
d = diag(H);
cvx_begin gp
variables p(m,1) t
maximize t
subject to
t*((H-diag(d))'*p + noise_pow) ./ (d.*p) <= 1
p <= pmax
p >= 0
cvx_end

p
```

```
Status: Solved
Optimal value (cvx_optval): +1.35762
p =

    0.6863
    0.9367
    0.9367
    0.6863
%=========================================
```

4.5.4 Power Dispatching in Interconnected Microgrids

Distributed power generation is a key concept in the next generation of power systems [15]. Small-scale generators and renewable energy sources can be collected together to supply heterogeneous power demands. Because of its distributed structure, the system is more reliable in terms of maintenance and services, as well as is more flexible in using renewable energy sources.

Integrating distributed power units into a framework, using information, communication and smart technologies, is known as a smart power grid [16]. This grid, characterized by power flexibility and reliability, enables the incorporation of various components such as renewable power resources and distributed micro-generators.

A unit of the grid, known as a microgrid (MG), is a group of small generators and loads connected to the grid in multiple points. As a considerable capability, each MG can operate in islanded and grid-connected operation modes. In the grid-connected operation mode, as in Fig. 4.11, each MG interacts with other MGs to establish a set of regulating power flows in the grid [18, 19]. As a result of this regulation, a grid-wide balance between supply and demand is provided, which results in a low operational cost.

Consider a distributed power grid consisting of a set $\mathcal{N} \triangleq \{i : i = 1, \dots, n\}$ of MGs interconnected through a power and data transmission infrastructure. Power demand within each MG_i is assumed to

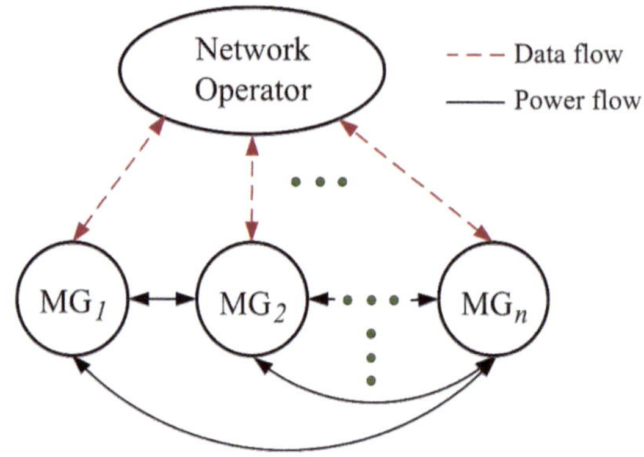

Fig. 4.11 Interconnected MGs [17]

be γ_i. Moreover, it is assumed that total power generated in MG$_i$ must not exceed σ_i. The objective of the network operator is to establish a set $\mathbf{X} = \{x_{ij}\}$ of power flows among MGs in order to minimize a cost function of the system-wide power generation. x_{ij} is the amount of power transmitted from MG$_i$ to MG$_j$. Under the assumption of a quadratic cost function for power generation at each MG, the optimization problem is formulated as

$$\min_{\mathbf{X}} \sum_{i=1}^{n} \left(\sum_{j=1}^{n} x_{ij} \right)^2 \tag{4.33a}$$

$$\text{subject to} \sum_{i=1}^{n} x_{ij} = \gamma_j, \qquad j = 1, \ldots, n \tag{4.33b}$$

$$\sum_{j=1}^{n} x_{ij} \leq \sigma_i, \qquad i = 1, \ldots, n \tag{4.33c}$$

$$x_{ij} \geq 0. \qquad i, j = 1, \ldots, n \tag{4.33d}$$

Constraint (4.33b) satisfies the supply–demand balance in each MG$_j$ and constraint (4.33c) restricts the power supplied by each MG$_i$ to its maximum power generation level. This convex problem is solved for three MGs with the same σ_i's but different γ_i's, i.e., demands, as in the following:

```
%============================================
d = [1; 2; 3]; % demand
s = [3; 3; 3]; % supply
n = 3;
b = ones(n, 1);

cvx_begin
variables x(n, n)
minimize sum(x(1, :))^2 + sum(x(2, :))^2
subject to
x'*b == d
x*b <= s
x >= 0
cvx_end
```

```
Status:  Solved
Optimal value (cvx_optval): +4.5
>> x

x =

    0.3051    0.5253    0.6696
    0.3051    0.5253    0.6696
    0.3899    0.9493    1.6608

%===============================================
```

4.6 Summary

In this chapter, we introduced the fundamentals of convex programming including convex sets, convex functions, and convex problem. In particular, the first- and second-order conditions on convex functions are presented and the optimality condition has been shown to be both necessary and sufficient for convex problems. Furthermore, geometric programming as a special optimization that can be transformed into a convex form has also been introduced. A number of application examples in electrical engineering with convex models have been introduced and their numerical solutions using CVX have been presented. In the next chapter, the approaches and algorithms to solve convex problems will be presented.

4.7 Problems

Problem 1 For each of the following functions determine whether it is convex, concave, or neither one. In case of neither convex or concave, find a transformation to make the function convex.

(a) $f(\mathbf{x}) = \frac{1}{x_1 x_2}$ on $\mathbb{R}^2_{++}$.
(b) $f(\mathbf{x}) = \frac{x_1}{x_2^2}$ on $\mathbb{R}^2_{++}$.

Problem 2 Consider the optimization problem

$$\begin{aligned} \min \quad & x_1^2 + x_2^2 \\ \text{subject to} \quad & x_2 \geq -x_1 + 2 \\ & x_2 \geq x_1 \\ & x_1 \geq 0, x_2 \geq 0. \end{aligned}$$

(a) Plot the feasible region and check out if this region is convex.
(b) Determine whether the problem is convex and find the optimal value.

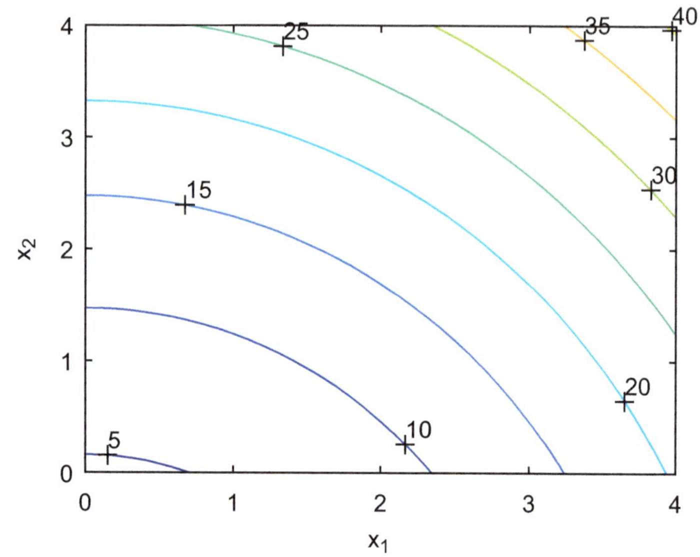

Fig. 4.12 Level set of $f(\mathbf{x}) = x_1^2 + 0.5x_2^2 + 3x_2 + 4.5$

Problem 3 Figure 4.12 is the level set diagram of function $f(\mathbf{x}) = x_1^2 + 0.5x_2^2 + 3x_2 + 4.5$.

(a) Using the level set diagram, show that the function is convex.
(b) Derive the second-order derivative of $f(\mathbf{x})$ and verify your answer in part (b).
(c) Now consider the problem

$$\begin{aligned} \text{minimize} \quad & f(\mathbf{x}) \\ \text{subject to} \quad & x_1 \geq 0,\ x_2 \geq 0. \end{aligned}$$

Show that $x^* = [0\ 0]^T$ satisfies the first-order optimality condition.

Problem 4 Show that exponential function $f(x) = e^{ax}\ \forall x, a \in \mathbb{R}$ is convex and its inverse logarithm function $f(x) = \ln x$ for $x > 0$ is concave.

Problem 5 Show that the power function $f(x) = x^\alpha$ is convex for $x > 0,\ \alpha \geq 1$ or $\alpha \leq 0$ and is concave for $x > 0, 0 \leq \alpha \leq 1$.

Problem 6 Show that if $f_1, f_2, \ldots, f_n$ are convex functions, then $f(\mathbf{x}) = \max\{f_1(\mathbf{x}), \ldots, f_n(\mathbf{x})\}$ is also convex.

Problem 7 Consider a random vector $\mathbf{x} \in \mathbb{R}^n$ and convex function f. Show that

$$f(\mathbb{E}[\mathbf{x}]) \leq \mathbb{E}[f(\mathbf{x})]$$

where $\mathbb{E}[.]$ is the expected value function. This inequality is known as Jensen's inequality.

Problem 8 Show that minimizing the maximum of a finite number of posynomials, i.e., $F(\mathbf{x}) = \max_i\{f_i(\mathbf{x})\}$ is a geometric programming.

References

1. S. Boyd, L. Vandenberghe, *Convex Optimization* (Cambridge, Cambridge University Press, 2004)
2. H. Hindi, A tutorial on convex optimization ii: duality and interior point methods, in *Proceedings of the American Control Conference*, 2006
3. H. Hindi, A tutorial on convex optimization, in *Proceedings of the American Control Conference*, 2004
4. Y. Nesterov, A. Nemirovsky, *Interior Point Polynomial Algorithms in Convex Programming* (SIAM, Philadelphia, 1994)
5. M.S. Bazaraa, J.J. Jarvis, H.D. Sherali, *Linear Programming and Network Flows* (Wiley, Hoboken, 2009)
6. S. Boyda, S.J. Kim, L. Vandenberghe, A. Hassibi, A tutorial on geometric programming. Optim. Eng. **8**(1), 67–127 (2007)
7. K.O. Kortanek, X. Xu, Y. Ye, An infeasible interior-point algorithm for solving primal and dual geometric programs. Math. Program. **76**(1), 155–181 (1996)
8. Z.-Q. Luo, W. Yu, An introduction to convex optimization for communications and signal processing. IEEE J. Sel. Areas Commun. **24**, 1426–1438 (2006)
9. J.A. Taylor, *Convex Optimization of Power Systems* (Cambridge University Press, Cambridge, 2015)
10. M. Grant, S. Boyd, CVX: Matlab software for disciplined convex programming, version 2.1 (2014). http://cvxr.com/cvx
11. GNU linear programming kit, Version 4.45. http://www.gnu.org/software/glpk
12. MOSEK, Mosek optimization toolbox (2002). www.mosek.com
13. W. Stanczak, M. Wiczanowski, H. Boche, *Fundamentals of Resource Allocation in Wireless Networks: Theory and Algorithms* (Springer, Berlin, 2008)
14. S. Kandukuri, S. Boyd, Optimal power control in interference limited fading wireless channels with outage probability specifications. IEEE Trans. Wirel. Commun. **1**, 46–55 (2002)
15. H. Bevrani, B. Francoist, T. Ise, *Microgrid Dynamics and Control* (Wiley, Hoboken, 2017)
16. H. Bevrani, M. Watanabe, Y. Mitani, *Power System Monitoring and Control* (IEEE-Wiley, Hoboken, 2014)
17. M. Fathi, H. Bevrani, Regulating power management in interconnected microgrids. J. Renew. Sust. Energy **9**, 055502 (2017)
18. M. Fathi, H. Bevrani, Adaptive energy consumption scheduling for connected microgrids under demand uncertainty. IEEE Trans. Power Deliv. **28**, 1576–1583 (2013)
19. M. Fathi, H. Bevrani, Statistical cooperative power dispatching in interconnected microgrids. IEEE Trans. Sust. Energy **4**, 586–593 (2013)

5 Duality

Abstract

One common approach towards solving an optimization problem, in general, is to transfer it from the primal domain into a dual domain. This is sometimes of great advantage as the problem in the dual domain is simple enough to solve. In this chapter, this transformation is introduced and, based on the achievements from the dual domain, Karush–Kuhn–Tucker (KKT) conditions are derived to find optimal solutions in general optimization problems. Moreover, based on KKT conditions, Lagrangian algorithm is introduced to be implemented as an iterative search method to solve convex problems. Finally, some application examples in electrical engineering are given, accordingly.

Keywords

Duality · Dual decomposition · KKT conditions · Lagrangian algorithm · Sensitivity analysis

5.1 Lagrangian Function

Recall that the standard optimization problem, not necessarily convex, is given in the form of

$$\min_{\mathbf{x}} \; f_0(\mathbf{x}) \tag{5.1a}$$

$$\text{subject to} \; f_i(\mathbf{x}) \leq 0, \quad i = 1, \ldots, m \tag{5.1b}$$

$$h_i(\mathbf{x}) = 0, \quad i = 1, \ldots, p \tag{5.1c}$$

with domain $D = \cap_{i=0}^{m} D_{f_i} \cap_{i=1}^{p} D_{h_i}$. Let us consider this problem as a primal problem and define its corresponding *Lagrangian function* and *Lagrange multipliers* in Definition 1.

M. Fathi, H. Bevrani, *Optimization in Electrical Engineering*,
https://doi.org/10.1007/978-3-030-05309-3_5

Definition 1 Associated with each optimization problem in (5.1), the function $L : D \times \mathbb{R}^m \times \mathbb{R}^p \rightarrow \mathbb{R}$ in the form of

$$L(\mathbf{x}, \boldsymbol{\lambda}, \boldsymbol{\nu}) = f_0(\mathbf{x}) + \sum_{i=1}^{m} \lambda_i f_i(\mathbf{x}) + \sum_{i=1}^{p} \nu_i h_i(\mathbf{x}) \tag{5.2}$$

is called the *Lagrangian function*. Coefficients $\boldsymbol{\lambda} = \{\lambda_i\}_{i=1}^m$ and $\boldsymbol{\nu} = \{\nu_i\}_{i=1}^p$ are called Lagrange multipliers. Each λ_i is the Lagrange multiplier associated with $f_i(\mathbf{x}) \leq 0$ and each ν_i is the Lagrange multiplier associated with $h_i(\mathbf{x}) = 0$.

Indeed, Lagrangian function is a weighted sum of objective and constraint functions.

5.2 Dual Problem

Now, consider the function $g : \mathbb{R}^m \times \mathbb{R}^p \rightarrow \mathbb{R}$ defined as

$$g(\boldsymbol{\lambda}, \boldsymbol{\nu}) = \inf_{\mathbf{x} \in D} L(\mathbf{x}, \boldsymbol{\lambda}, \boldsymbol{\nu}). \tag{5.3}$$

This function is called the *dual function* associated with the so-called primal problem in (5.1). g is the pointwise minimum of a set of affine functions in $(\boldsymbol{\lambda}, \boldsymbol{\nu})$. Therefore, it is a concave function even though the primal problem is not convex, as discussed in Chap. 4.

Deriving a dual function for an optimization problem is of high importance as we can take advantage of beneficial Theorem 1.

Theorem 1 *Let p^* be the optimal value of the objective function in the standard optimization problem. Then, the dual function with all $\lambda_i \geq 0$ is the lower bound of p^*. In other words, if $\lambda_i \geq 0$ for all i, then $g(\boldsymbol{\lambda}, \boldsymbol{\nu}) \leq p^*$.*

Proof Suppose $\tilde{\mathbf{x}}$ be a feasible point of problem (5.1). If $\lambda_i \geq 0$ for all i, we claim

$$g(\boldsymbol{\lambda}, \boldsymbol{\nu}) = \inf_{\mathbf{x} \in D} L(\mathbf{x}, \boldsymbol{\lambda}, \boldsymbol{\nu}) \leq L(\tilde{\mathbf{x}}, \boldsymbol{\lambda}, \boldsymbol{\nu}) \leq f_0(\tilde{\mathbf{x}}).$$

The first inequality is due to the fact that the inf operation returns the minimum function value. The second inequality occurs because $f_i(\tilde{\mathbf{x}}) \leq 0$ and $h_i(\tilde{\mathbf{x}}) = 0$ as a result of the feasibility of $\tilde{\mathbf{x}}$. Since the optimal point $\mathbf{x}^*$ with function value $p^* = f(\mathbf{x}^*)$ is also a feasible point, it is concluded that $g(\boldsymbol{\lambda}, \boldsymbol{\nu}) \leq p^*$. In other words, the dual function is a lower bound of the optimal value. □

Following the lower bound property in Theorem 1, the best lower bound can be derived using a problem called *dual problem* in the form of

$$\max_{\boldsymbol{\lambda}, \boldsymbol{\nu}} \ g(\boldsymbol{\lambda}, \boldsymbol{\nu}) \tag{5.4a}$$

$$\text{subject to } \lambda_i \geq 0, \ i = 1, \ldots, m. \tag{5.4b}$$

Recall that $g(\boldsymbol{\lambda}, \boldsymbol{\nu})$ is concave, so the dual problem is always a convex problem even for non-convex problems in the primal domain. This problem can be used to find optimal Lagrange multipliers in the dual domain.

Assume that the optimal value of the dual problem has been derived as d^*, i.e., the best lower bound on p^* that can be achieved from the dual function. We define the *duality gap* as the difference between the optimal values of the primal and dual problems, i.e., $p^* - d^*$. As a result of the mentioned lower bound property in Theorem 1, $d^* \leq p^*$ and therefore the duality gap is always non-negative. In particular, if duality gap is zero, i.e., $d^* = p^*$, we say that *strong duality* holds or there exist a zero duality gap. Otherwise, there exist a *weak duality*, i.e., $d^* < p^*$.

Note that the strong duality does not hold in general. However, it usually holds for convex problems. For more information, refer to Slater's constraint qualification in the literature [1].

Dual problem derivation is more explained by the following two examples.

Example 1 Find the dual problem for the least norm solution of linear equations in the form of

$$\min \quad \mathbf{x}^T\mathbf{x}$$
$$\text{subject to } \mathbf{A}\mathbf{x} = \mathbf{b}.$$

Solution The Lagrangian function is written as $L(\mathbf{x}, \boldsymbol{\nu}) = \mathbf{x}^T\mathbf{x} + \boldsymbol{\nu}^T(\mathbf{A}\mathbf{x} - \mathbf{b})$. Therefore, the dual function is derived as $g(\boldsymbol{\nu}) = \inf_{\mathbf{x}\in D} L(\mathbf{x}, \boldsymbol{\nu})$. The solution can be done using

$$\nabla_{\mathbf{x}} L(\mathbf{x}, \boldsymbol{\nu}) = 2\mathbf{x} + \mathbf{A}^T\boldsymbol{\nu} = 0 \Rightarrow \mathbf{x} = -\frac{1}{2}\mathbf{A}^T\boldsymbol{\nu}$$
$$\Rightarrow g(\boldsymbol{\nu}) = L(\mathbf{x}, \boldsymbol{\nu}) = -\frac{1}{4}\boldsymbol{\nu}^T\mathbf{A}\mathbf{A}^T\boldsymbol{\nu} - b^T\boldsymbol{\nu}.$$

Then, the dual problem is to maximize $g(\boldsymbol{\nu})$. It can be verified that $g(\boldsymbol{\nu})$ is a concave function as $-\mathbf{A}\mathbf{A}^T$ is negative semi-definite. □

Example 2 Find the dual problem of the following quadratic problem. Assume $\mathbf{P}$ is positive semi-definite.

$$\min \quad \mathbf{x}^T\mathbf{P}\mathbf{x}$$
$$\text{subject to } \mathbf{A}\mathbf{x} \leq \mathbf{b}$$

Solution The Lagrangian is $L(\mathbf{x}, \boldsymbol{\lambda}) = \mathbf{x}^T\mathbf{P}\mathbf{x} + \boldsymbol{\lambda}^T(\mathbf{A}\mathbf{x} - \mathbf{b})$ and therefore the dual function is derived as $g(\boldsymbol{\lambda}) = \inf_{\mathbf{x}\in D} L(\mathbf{x}, \boldsymbol{\lambda})$. The solution can be done using

$$\nabla_{\mathbf{x}} L(\mathbf{x}, \boldsymbol{\lambda}) = 2\mathbf{P}\mathbf{x} + \mathbf{A}^T \boldsymbol{\lambda} = 0$$
$$\Rightarrow \mathbf{x} = -\frac{1}{2}\mathbf{P}^{-1}\mathbf{A}^T\boldsymbol{\lambda}$$
$$\Rightarrow g(\boldsymbol{\lambda}) = -\frac{1}{4}\boldsymbol{\lambda}^T\mathbf{A}\mathbf{P}^{-1}\mathbf{A}^T\boldsymbol{\lambda} - b^T\boldsymbol{\lambda}.$$

Consequently, the dual problem is to maximize $g(\boldsymbol{\lambda})$. □

5.3 Karush–Kuhn–Tucker Conditions

In mathematical programming, Karush–Kuhn–Tucker (KKT) conditions were originally named after Harold W. Kuhn and Albert W. Tucker first published the conditions in 1951 [2]. These conditions reduce the solution of a *zero duality gap* problem into a set of equations and inequalities that can be solved analytically or numerically.

Prior to introducing KKT conditions, let us discuss two important properties derived from strong duality. In a standard optimization problem, in which strong duality holds, suppose that $\mathbf{x}^*$ be primal optimal and $(\boldsymbol{\lambda}^*, \boldsymbol{\nu}^*)$ be dual optimal. Now, consider the statements in

$$f_0(\mathbf{x}^*) = g(\boldsymbol{\lambda}^*, \boldsymbol{\nu}^*) = \inf_{\mathbf{x}\in D}\left(f_0(\mathbf{x}) + \sum_{i=1}^{m}\lambda_i^* f_i(\mathbf{x}) + \sum_{i=1}^{p}\nu_i^* h_i(\mathbf{x})\right) \tag{5.5}$$
$$\le f_0(\mathbf{x}^*) + \sum_{i=1}^{m}\lambda_i^* f_i(\mathbf{x}^*) + \sum_{i=1}^{p}\nu_i^* h_i(\mathbf{x}^*) \tag{5.6}$$
$$\le f_0(\mathbf{x}^*). \tag{5.7}$$

In this statement, the concept of strong duality, dual function, and Lagrangian function has been utilized together. The inequality in (5.6) is due to the fact that the minimum value of a function is always less than or equal to the function for each possible argument in the domain, $\mathbf{x}^*$ in this case. Moreover, the inequality in (5.7) comes from the feasibility of $\mathbf{x}^*$ and $\boldsymbol{\lambda}^*$ that implies $\lambda_i^* f_i(\mathbf{x}^*) \le 0$ and $h_i(\mathbf{x}^*) = 0$ for all i.

From (5.5)–(5.7), we conclude that the two inequalities in (5.6) and (5.7) can be replaced with equality. Considering this derivation, following statements are concluded:

1. From the first one in (5.6), we derive $\inf_{\mathbf{x}\in D} L(\mathbf{x}, \boldsymbol{\lambda}^*, \boldsymbol{\nu}^*) = L(\mathbf{x}^*, \boldsymbol{\lambda}^*, \boldsymbol{\nu}^*)$. In other words, $\mathbf{x}^*$ as an optimal primal variable minimizes the Lagrangian function $L(\mathbf{x}, \boldsymbol{\lambda}^*, \boldsymbol{\nu}^*)$ as well.
2. From the second one in (5.7), it is concluded that $\lambda_i^* f_i(\mathbf{x}^*) = 0$ for $i = 1, \ldots, m$. This is known as *complementary slackness* property. In other words, for each $f_i(\mathbf{x}^*)$ and λ_i^*, we can say that either

$$\lambda_i^* > 0 \Rightarrow f_i(\mathbf{x}^*) = 0 \tag{5.8}$$

or

$$f_i(\mathbf{x}^*) < 0 \Rightarrow \lambda_i^* = 0. \tag{5.9}$$

Considering the aforementioned derivations, KKT conditions are stated in Theorem 2.

Theorem 2 *In any optimization problem with differentiable objective and constraint functions for which strong duality holds, any pair of primal* $\mathbf{x}^*$ *and dual* $(\boldsymbol{\lambda}^*, \boldsymbol{\nu}^*)$ *optimal points must satisfy the KKT conditions as in*

1. $f_i(\mathbf{x}^*) \leq 0, \quad i = 1, \ldots, m.$
2. $h_i(\mathbf{x}^*) = 0, \quad i = 1, \ldots, p.$
3. $\lambda_i^* \geq 0, \quad i = 1, \ldots, m.$
4. $\lambda_i^* f_i(\mathbf{x}^*) = 0, \quad i = 1, \ldots, m.$
5. $\nabla_{\mathbf{x}} L(\mathbf{x}^*, \boldsymbol{\lambda}^*, \boldsymbol{\nu}^*) = \nabla f_0(\mathbf{x}^*)$
   $+ \sum_{i=1}^{m} \lambda_i^* \nabla f_i(\mathbf{x}^*) + \sum_{i=1}^{p} \nu_i^* \nabla h_i(\mathbf{x}^*) = 0.$

The first three conditions are from the fact that $\mathbf{x}^*$ and $\boldsymbol{\lambda}^*$ are feasible. The forth condition is from the complementary slackness property. Finally, the last condition is due to the fact that $\mathbf{x}^*$ minimizes $L(\mathbf{x}, \boldsymbol{\lambda}^*, \boldsymbol{\nu}^*)$ function and therefore the gradient of Lagrangian function with respect to $\mathbf{x}$ is zero at $\left(\mathbf{x}^*, \boldsymbol{\lambda}^*, \boldsymbol{\nu}^*\right)$.

It is noteworthy that the aforementioned KKT conditions are only necessary for optimal points in general optimization problems. However, in the case of convex problems, it is shown in Theorem 3 that KKT conditions are both necessary and sufficient conditions. In other words, these conditions result in optimal primal and dual solutions, and zero duality gap.

Theorem 3 *In a convex optimization problem with differentiable objective and constraint functions, if any* $\left(\tilde{\mathbf{x}}, \tilde{\boldsymbol{\lambda}}, \tilde{\boldsymbol{\nu}}\right)$ *satisfies KKT conditions, then* $\tilde{\mathbf{x}}$ *is primal optimal,* $(\tilde{\boldsymbol{\lambda}}, \tilde{\boldsymbol{\nu}})$ *is dual optimal, and strong duality holds.*

Proof If any $\left(\tilde{\mathbf{x}}, \tilde{\boldsymbol{\lambda}}, \tilde{\boldsymbol{\nu}}\right)$ satisfies KKT conditions, the first three conditions in Theorem 2 state that $\tilde{\mathbf{x}}$ and $\left(\tilde{\boldsymbol{\lambda}}, \tilde{\boldsymbol{\nu}}\right)$ are primal and dual feasible. Besides, $\tilde{\lambda}_i \geq 0$ for all i in condition 3 implies that $L(\mathbf{x}, \tilde{\boldsymbol{\lambda}}, \tilde{\boldsymbol{\nu}})$ is also a convex function in $\mathbf{x}$. Therefore, $\nabla_{\mathbf{x}} L(\tilde{\mathbf{x}}, \tilde{\boldsymbol{\lambda}}, \tilde{\boldsymbol{\nu}}) = 0$ in condition 5 states that $\tilde{\mathbf{x}}$ minimizes $L(\mathbf{x}, \tilde{\boldsymbol{\lambda}}, \tilde{\boldsymbol{\nu}})$. In other words,

$$g(\tilde{\boldsymbol{\lambda}}, \tilde{\boldsymbol{\nu}}) = \inf_{\mathbf{x} \in D} L(\mathbf{x}, \tilde{\boldsymbol{\lambda}}, \tilde{\boldsymbol{\nu}}) = L(\tilde{\mathbf{x}}, \tilde{\boldsymbol{\lambda}}, \tilde{\boldsymbol{\nu}}).$$

Moreover, since $\tilde{\lambda}_i f_i(\tilde{\mathbf{x}}) = 0$ and $h_i(\tilde{\mathbf{x}}) = 0$ for all i, we can also say that $L(\tilde{\mathbf{x}}, \tilde{\boldsymbol{\lambda}}, \tilde{\boldsymbol{\nu}}) = f_0(\tilde{\mathbf{x}})$ and consequently

$$g(\tilde{\boldsymbol{\lambda}}, \tilde{\boldsymbol{\nu}}) = f_0(\tilde{\mathbf{x}}).$$

This derivation implies that the pair of $\tilde{\mathbf{x}}$ and $(\tilde{\boldsymbol{\lambda}}, \tilde{\boldsymbol{\nu}})$ provides zero duality gap and consequently they are primal and dual optimal. □

Following the results of Theorems 2 and 3, it is concluded that for a convex problem with zero duality gap, KKT conditions are both necessary and sufficient conditions. Based on these conclusions, many algorithms exist for solving convex optimization problems using KKT conditions in the literature [3, 4]. One of these algorithms, known as the Lagrangian algorithm, is stated in Sect. 5.4.

Example 3 Solve the least norm problem in Example 1.

Solution The problem is convex. Therefore, $\mathbf{x} = -(1/2)\mathbf{A}^T\boldsymbol{\nu}$ derived from $\nabla_{\mathbf{x}}L(\mathbf{x}, \boldsymbol{\nu}) = 0$ could also be primal optimal if $\boldsymbol{\nu}$ is substituted with the dual optimal value. Since the derived dual problem in the solution of Example 1 is an unconstrained problem, we can use $\nabla_{\boldsymbol{\nu}}g(\boldsymbol{\nu}) = 0$ to find the optimal Lagrange multiplier $\boldsymbol{\nu}^*$ and then $\mathbf{x}^*$ as

$$\nabla_{\boldsymbol{\nu}}g(\boldsymbol{\nu}) = 0 \Rightarrow \boldsymbol{\nu}^* = -2(\mathbf{A}\mathbf{A}^T)^{-1}\mathbf{b} \Rightarrow \mathbf{x}^* = \mathbf{A}^T(\mathbf{A}\mathbf{A}^T)^{-1}\mathbf{b}. \qquad \square$$

Example 4 Find the KKT conditions for the following problem with $\mathbf{P} \succeq 0$.

$$\begin{aligned} \min\ & \tfrac{1}{2}\mathbf{x}^T\mathbf{P}\mathbf{x} + \mathbf{q}^T\mathbf{x} + r \\ \text{subject to}\quad & \mathbf{A}\mathbf{x} = \mathbf{b} \end{aligned}$$

Solution This is a quadratic convex problem with only affine constrains. The Lagrangian function and its gradient are $L(\mathbf{x}, \boldsymbol{\nu}) = \frac{1}{2}\mathbf{x}^T\mathbf{P}\mathbf{x} + \mathbf{q}^T\mathbf{x} + r + \boldsymbol{\nu}^T(\mathbf{A}\mathbf{x} - \mathbf{b})$ and $\nabla_{\mathbf{x}}L(\mathbf{x}, \boldsymbol{\nu}) = \mathbf{P}\mathbf{x} + \mathbf{q} + \mathbf{A}^T\boldsymbol{\nu}$, respectively. Therefore, the KKT conditions that are necessary and sufficient conditions are derived as

$$\begin{cases} \mathbf{A}\mathbf{x} - \mathbf{b} = 0 \\ \mathbf{P}\mathbf{x} + \mathbf{q} + \mathbf{A}^T\boldsymbol{\nu} = 0 \end{cases} \Rightarrow \begin{bmatrix} \mathbf{P} & \mathbf{A}^T \\ \mathbf{A} & 0 \end{bmatrix} \begin{bmatrix} \mathbf{x} \\ \boldsymbol{\nu} \end{bmatrix} = \begin{bmatrix} -\mathbf{q} \\ \mathbf{b} \end{bmatrix}.$$

As derived, the problem solution is reduced to solve a set of linear equations. □

5.4 Lagrangian Algorithm

So far, we have transformed the optimization problem in the primal domain (5.1) into an equivalent problem in the dual domain (5.4). As highlighted in Sects. 5.2, 5.3, finding the optimal Lagrange multipliers $(\boldsymbol{\lambda}^*, \boldsymbol{\nu}^*)$ in the dual domain is of great help to find the optimal solution $\mathbf{x}^*$. This section presents Lagrangian algorithm which uses both primal and dual domains jointly to reach the optimal solution.

In Chap. 3, we have discussed a number of iterative search methods to solve unconstrained optimization problems. Here, we can take advantage of these methods to solve an optimization problem jointly in the primal and dual domains. This is based on the following facts. First, the optimal $\mathbf{x}^*$ that minimizes $f(\mathbf{x})$ is also the minimizer of the Lagrangian function $L(\mathbf{x}, \boldsymbol{\lambda}^*, \boldsymbol{\nu}^*)$. In other words, $\mathbf{x}^*$ solves

$$\min_{\mathbf{x}} \quad L(\mathbf{x}, \boldsymbol{\lambda}^*, \boldsymbol{\nu}^*) \tag{5.10}$$

as well, which is an unconstrained problem. Second, the dual problem,

$$\max_{\boldsymbol{\lambda}, \boldsymbol{\nu}} \ g(\boldsymbol{\lambda}, \boldsymbol{\nu}) \tag{5.11a}$$

$$\text{subject to } \lambda_i \geq 0, \quad i = 1, \ldots, m \tag{5.11b}$$

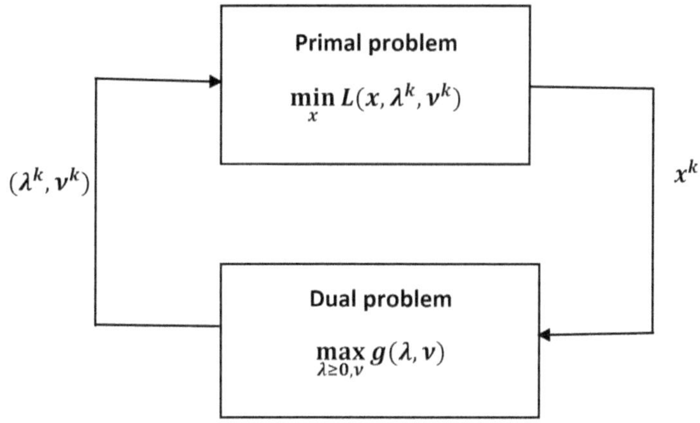

Fig. 5.1 Lagrangian algorithm

is mostly an unconstrained problem. Even if not, due to the constraint of (5.11b), we can relax this constraint and then take advantage of projection methods later to apply this constraint to the derived solution.

Considering the aforementioned points, the Lagrangian algorithm for constrained convex problems is depicted in Fig. 5.1, where iterative search methods are deployed in both primal and dual domains together. The steps are taken as in the following:

1. First, supply starting points $\mathbf{x}^0$, $\boldsymbol{\lambda}^0$, $\boldsymbol{\nu}^0$ and a stopping criteria for each variable.
2. Assuming given points $\mathbf{x}^k$, $\boldsymbol{\lambda}^k$, $\boldsymbol{\nu}^k$ at iteration k, the gradients of $L(\mathbf{x}, \boldsymbol{\lambda}, \boldsymbol{\nu})$ and $g(\boldsymbol{\lambda}, \boldsymbol{\nu})$ at $\mathbf{x}^k$, $\boldsymbol{\lambda}^k$, $\boldsymbol{\nu}^k$ are computed as

$$\nabla_{\mathbf{x}} L(\mathbf{x}^k, \boldsymbol{\lambda}^k, \boldsymbol{\nu}^k) = \nabla f_0(\mathbf{x}^k) + \sum_{i=1}^{m} \lambda_i^k \nabla f_i(\mathbf{x}^k) + \sum_{i=1}^{p} \nu_i^k \nabla h_i(\mathbf{x}^k)$$

$$\frac{\partial g(\boldsymbol{\lambda}^k, \boldsymbol{\nu}^k)}{\partial \lambda_i} = f_i(\mathbf{x}^k), \quad i = 1, 2, \ldots, m$$

$$\frac{\partial g(\boldsymbol{\lambda}^k, \boldsymbol{\nu}^k)}{\partial \nu_i} = h_i(\mathbf{x}^k), \quad i = 1, 2, \ldots, p.$$

These gradients are used as search directions with appropriate step sizes in iterative methods.
3. The variables in the next iterate $k + 1$ are updated numerically as

$$\mathbf{x}^{k+1} = \mathbf{x}^k - \alpha_k \left(\nabla f_0(\mathbf{x}^k) + \sum_{i=1}^{m} \lambda_i^k \nabla f_i(\mathbf{x}^k) + \sum_{i=1}^{p} \nu_i^k \nabla h_i(\mathbf{x}^k) \right) \tag{5.12}$$

$$\lambda_i^{k+1} = \left(\lambda_i^k + \beta_k f_i(\mathbf{x}^k) \right)^+, \quad i = 1, 2, \ldots, m \tag{5.13}$$

$$\nu_i^{k+1} = \nu_i^k + \gamma_k h_i(\mathbf{x}^k), \quad i = 1, 2, \ldots, p \tag{5.14}$$

where $\alpha_k, \beta_k, \gamma_k$ are step sizes at iterate k and $(.)^+ \triangleq \max(., 0)$. Since dual variables are to maximize the dual problem, search directions are in the direction of gradients, i.e., gradients are considered with positive signs for dual variables in (5.13) and (5.14). Moreover, the constraints $\lambda_i \geq 0$ have been applied the operation of $(.)^+$ in (5.13), indicating the projection method noted earlier in this section.

Steps 2 and 3 repeat until the satisfaction of a termination criterion, e.g., the convergence of $\mathbf{x}$, $\boldsymbol{\lambda}$, and $\boldsymbol{\nu}$ in Step 3.

The examples on how to apply the Lagrangian algorithm to solve convex optimization problems are addressed in Sect. 5.5.

5.5 Application Examples

5.5.1 Power Control in Cellular Communication

Consider a wireless communication system, as in Fig. 5.2, consisting of a base station as the transmitter and n users as receivers with channel gains $\{h_i\}_{i=1}^n$. Depending on environmental conditions, channel gains are time-varying following certain statistical characteristics. To address the impairment effects of time-varying channel gains, the base station is to adopt transmit power p_i to provide transmit rate $r_i = \log_2(1 + \alpha_i h_i p_i)$ for each user i, where α_i is a positive scalar.

Now, assume that the base station wishes to maximize its sum transmit rate $\sum_{i=1}^n r_i$ to the users by adopting the transmit powers $\{p_i\}_{i=1}^n$. Moreover, assume that the total transmit power cannot be exceeded by $P_{\max}$ at the base station. Hence, the problem is to determine $\{p_i\}_{i=1}^n$ such that the system sum-rate is maximized. This objective then can be formulated as

$$\max_{\mathbf{p}} \sum_{i=1}^{n} \log_2(1 + \alpha_i h_i p_i) \tag{5.15a}$$

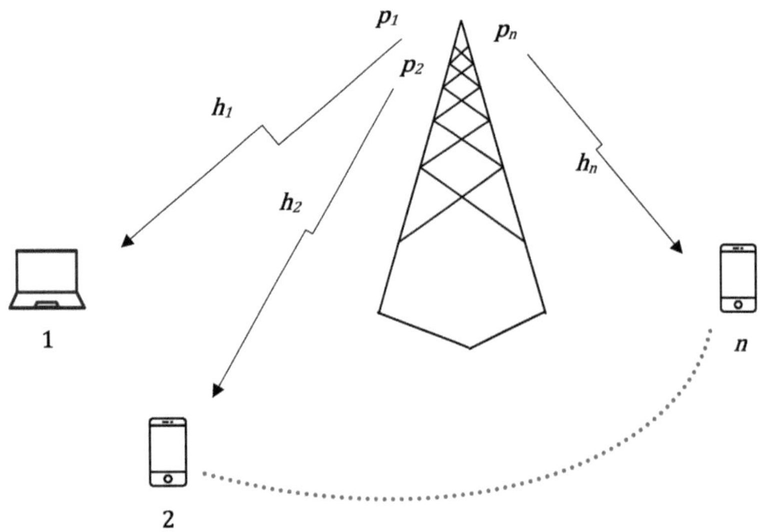

Fig. 5.2 Wireless power allocation

$$\text{subject to} \quad \sum_{i=1}^{n} p_i \leq P_{\max} \tag{5.15b}$$

$$p_i \geq 0, \quad i = 1, \ldots, n \tag{5.15c}$$

Logarithmic functions are concave and $(1+\alpha_i h_i p_i)$ is affine, thus the objective function is concave. The constraints are also affine and convex. Therefore, the problem is convex and can be rewritten as

$$\min_{\mathbf{p}} \quad -\sum_{i=1}^{n} \log_2(1 + \alpha_i h_i p_i) \tag{5.16a}$$

$$\text{subject to} \quad \sum_{i=1}^{n} p_i - P_{\max} \leq 0 \tag{5.16b}$$

$$-p_i \leq 0, \quad i = 1, \ldots, n \tag{5.16c}$$

For the time being, relax the constraint (5.16c) as it can be applied later to the solution. The Lagrangian and dual functions of problem (5.16) are written as

$$L(\mathbf{p}, \lambda) = -\sum_{i=1}^{n} \log_2(1+\alpha_i h_i p_i) + \lambda\left(\sum_{i=1}^{n} p_i - P_{\max}\right) \tag{5.17}$$

and $g(\lambda) = \inf_{\mathbf{p}\in D} L(\mathbf{p}, \lambda)$, respectively. Since the primal problem is convex, optimal transmit powers also minimize $L(\mathbf{p}, \lambda^*)$. Taking the gradient of this function, we derive

$$\nabla_{\mathbf{p}} L(\mathbf{p}, \lambda^*) = 0 \Rightarrow p_i^* = \left(\frac{1}{\lambda^* \ln 2} - \frac{1}{\alpha_i h_i}\right)^+ \tag{5.18}$$

for all i, where $(.)^+ \equiv \max(., 0)$ is applied to satisfy constraint (5.16c). Therefore, the dual function and dual problem are obtained as $g(\lambda) = L(\mathbf{p}^*, \lambda)$ and $\max_{\lambda\geq 0} g(\lambda)$, respectively. This solution is usually known as water-falling power allocation in the literature of wireless communication [5].

As a numerical example, consider a network of $n = 10$ users, channel gains $h = \{0.1, 0.2, \ldots, 1\}$, $\alpha_i = 1$ for all i, and $P_{\max} = 1$. Taking a numerical computation to substitute p_i^*s in (5.18) into $g(\lambda) = L(\mathbf{p}^*, \lambda)$, we plot the dual function in Fig. 5.3. As shown, this function is maximized for $\lambda^* \simeq 1$. Substituting in (5.18), the transmit powers are consequently computed as $p_i^* = 0$ for $i = 1, \ldots, 6$, $p_7^* = 0.0141$, $p_8^* = 0.1927$, $p_9^* = 0.3316$, and $p_{10}^* = 0.4427$.

In addition to this numerical solution, we also take advantage of the CVX solver in MATLAB to obtain the optimal transmit powers as in the following.

Fig. 5.3 Dual function

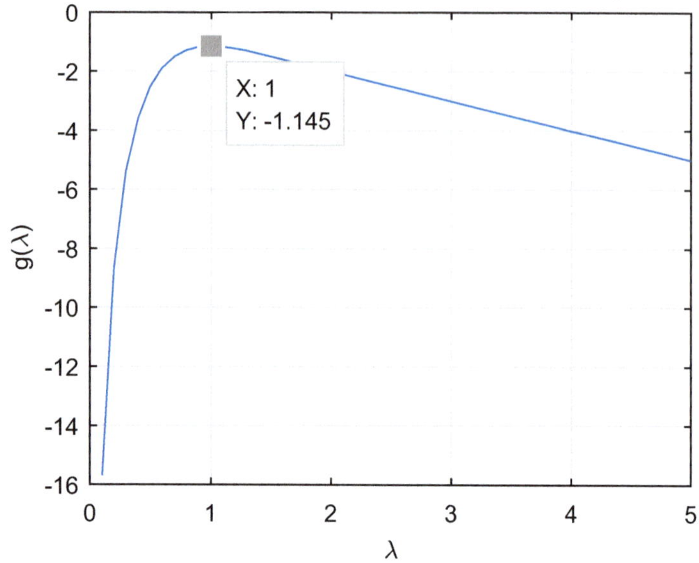


```
%========================================
clear all
h = 0.1:0.1:1; Pmax = 1;
cvx_begin
variable p(1,10)
dual variable lambda
maximize sum(log(1+h.*p)/log(2))
subject to
lambda: sum(p) <= Pmax
p >= 0
cvx_end
p
lambda
```

```
Optimal value (cvx_optval): 1.14543
p =
     0.0000    0.0000    0.0000    0.0000    0.0000    0.0000    0.0189    0.1974
0.3363    0.4474

lambda =
     0.9968
%========================================
```

These results mostly match the ones achieved using numerical solution. As a remarkable point, the Lagrange multiplier associated with each constraint can also be found in CVX using `dual variable` script and the constraint labeled by the variable name. As shown, the optimal Lagrange multiplier for constraint (5.16b) derived by CVX is $\lambda^* = 0.9968$.

As an alternative solution for (5.16), we take advantage of Lagrangian algorithm. Recall that taking $\nabla L_{\mathbf{p}}(\mathbf{p}, \lambda^*) = 0$ yields $p_i = \left(\frac{1}{\lambda^* \log(2)} - \frac{1}{h_i}\right)^+$ for all i. Under the assumption of a given dual variable λ, we can take advantage of this to derive transmit powers. Rather than plotting the dual function and finding the optimal dual variable, here, we employ an iterative search method. Taking the gradient of dual problem with respect to λ results in

$$\frac{\partial g}{\partial \lambda} = \sum_{i=1}^{n} p_i - P_{\max}.$$

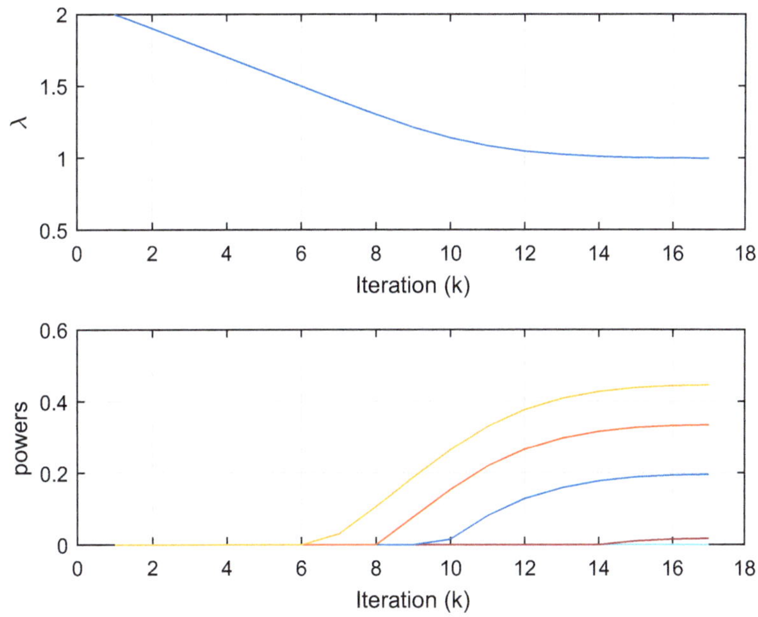

Fig. 5.4 Lagrange multiplier and powers

Then, the iterative search method can be formed as

$$\lambda^{k+1} = \left(\lambda^k + \beta^k \left(\sum_{i=1}^{n} p_i^k - P_{\max}\right)\right)^+$$

where β^k is the step size and p_i^k is the transmit power of user i at iteration k. Considering these derivations, dual variable λ and transmit powers are plotted in Fig. 5.4 versus search iterations. As shown, after a number of iterations, the curves converge to the same results in two previous solutions.

5.5.2 Cross-Layer Resource Allocation

Traditionally, the functionality of protocol stacks in communication networks is divided between different layers, based on OSI or TCP/IP reference models [6, 7]. The performance functionality of each layer is mostly independent from the other layers. Therefore, the performance optimization of each layer is done individually [8].

However, in the case of wireless networks, the performance of network layers is mostly coupled with each other [9]. For example, the power control and bandwidth allocation in the physical (PHY) layer determines link establishment throughout the network. This in turn impacts the channel assignment in the multiple access control (MAC) layer and the routes and queue management schemes adopted by the *network* layer. Following these resource allocation mechanisms, the end-to-end rate control and congestion control policies are accomplished in the *transport* layer. In other words, the performance optimization of a wireless network is required to be done jointly across the whole layers in the reference model. This is usually known as cross-layer design in the literature [10–12].

To illustrate the application of duality in the cross-layer design of wireless networks, consider a multi-hop wireless network, as in Fig. 5.5, consisting of L links, indexed by l, and S transmitters, indexed by s. The transmit rate of each transmitter s is indicated by r_s, which is routed and relayed

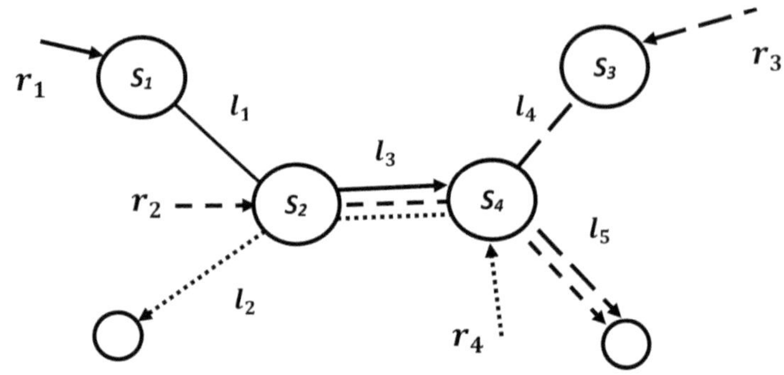

Fig. 5.5 A typical multi-hop network

through a path consisting of a set $\mathcal{L}(s)$ of wireless links. In the network depicted in Fig. 5.5, $\mathcal{L}(1) = \{l_1, l_3\}$, $\mathcal{L}(2) = \{l_3, l_5\}$, $\mathcal{L}(3) = \{l_4, l_5\}$, and $\mathcal{L}(4) = \{l_3, l_2\}$. The capacity of each link l is a function of the allocated resources, e.g., bandwidth and power in the MAC/PHY layer. The capacity is indicated by $c_l(\mathbf{w}, \mathbf{p})$, where $\mathbf{w}$ and $\mathbf{p}$ are allocated bandwidth and power vectors, respectively.

Assume that the network entity defines a utility function associated with each transmit rate r_s, denoted by $U_s(r_s)$. This is to measure the satisfaction level of user s from the received service r_s. The utility is usually a non-decreasing and concave function. The objective of the network entity is to maximize the sum of utilities throughout the network [13]. This is known as utility maximization problem and is formulated as

$$\max_{\mathbf{r},\mathbf{w},\mathbf{p}} \sum_{s=1}^{S} U_s(r_s) \tag{5.19a}$$

$$\text{subject to} \quad \sum_{s:l\in\mathcal{L}(s)} r_s \leq c_l(\mathbf{w}, \mathbf{p}), \qquad l = 1, \ldots, L \tag{5.19b}$$

$$r_s \geq 0, \qquad s = 1, \ldots, S \tag{5.19c}$$

where constraint (5.19b) states that traffic load on any link does not exceed the achievable capacity on that link.

In general, solving the utility maximization problem raised in (5.19) depends on the network topology, utility, and link capacity functions. By the way, it can be addressed using dual domain in general. The Lagrangian and dual functions can be written as

$$L(\mathbf{r}, \mathbf{w}, \mathbf{p}, \boldsymbol{\lambda}) = \sum_{s=1}^{S} U_s(r_s) - \sum_{l=1}^{L} \lambda_l \left(\sum_{s:l\in\mathcal{L}(s)} r_s - c_l(\mathbf{w}, \mathbf{p}) \right) \tag{5.20}$$

and

$$D(\boldsymbol{\lambda}) = \max_{\mathbf{r},\mathbf{w},\mathbf{p}} \quad L(\mathbf{r}, \mathbf{w}, \mathbf{p}, \boldsymbol{\lambda}) \tag{5.21}$$

respectively, where $\boldsymbol{\lambda} = \{\lambda_l \geq 0\}_{l=1}^{L}$ is the vector of Lagrange multipliers associated with constraint (5.19b). For a given $\boldsymbol{\lambda}$, the maximization in (5.21) to derive the dual function is done using

$$\max_{\mathbf{r},\mathbf{w},\mathbf{p}} \sum_{s=1}^{S} U_s(r_s) - \sum_{l=1}^{L} \lambda_l \left(\sum_{s:l\in\mathcal{L}(s)} r_s - c_l(\mathbf{w}, \mathbf{p}) \right)$$

which can be decomposed into two sub-problems as

$$\max_{\mathbf{r}} \sum_{s=1}^{S} U_s(r_s) - \sum_{s=1}^{S} \sum_{l \in \mathcal{L}(s)} \lambda_l r_s \tag{5.22a}$$

$$+ \max_{\mathbf{w},\mathbf{p}} \sum_{l=1}^{L} \lambda_l c_l(\mathbf{w}, \mathbf{p}). \tag{5.22b}$$

The former in (5.22a) is a problem of transmit rates $\mathbf{r_s}$'s that is accomplished in the transmitters by the transport layer and the latter in (5.22b) is a problem of resource allocation on the links that is done by MAC/PHY layers. Two sub-problems are only coupled with each other using Lagrange multipliers in the dual domain. Therefore, this is usually known as *dual decomposition* [14].

The Lagrange multipliers $\boldsymbol{\lambda}$ are obtained using the dual problem as

$$\min_{\boldsymbol{\lambda} \geq 0} \quad D(\boldsymbol{\lambda}). \tag{5.23}$$

Using iterative search methods, the solution is achieved as

$$\lambda_l^{k+1} = \left(\lambda_l^k + \sum_{s: l \in \mathcal{L}(s)} r_s^k - c_l(\mathbf{w}^k, \mathbf{p}^k) \right)^+ \tag{5.24}$$

for all l, where r_s^k, $\mathbf{w}^k$, and $\mathbf{p}^k$ are primal variables at iteration k.

As shown, the system-wide problem of utility maximization in (5.19) has been decoupled into a number of sub-problems to be done by the network protocol stack. In particular, it results in jointly optimal end-to-end congestion control and per link resource allocation in the network.

The approach of dual decomposition has usually been adopted in the literature to decompose a system-wide problem into smaller problems and to perform cross-layer design [15–17]. The resulting sub-problems, solved by individual layers, are coupled with each other using Lagrange multipliers. As a remarkable point, this decomposition gains the complexity reduction of the system-wide problem.

5.6 Sensitivity Analysis

Recall that the standard primal and associated dual problems have been introduced as

$$\min_{\mathbf{x}} \quad f_0(\mathbf{x}) \tag{5.25a}$$

$$\text{subject to} \quad f_i(\mathbf{x}) \leq 0, \quad i = 1, \dots, m \tag{5.25b}$$

$$h_i(\mathbf{x}) = 0, \quad i = 1, \dots, p \tag{5.25c}$$

and

$$\max_{\boldsymbol{\lambda},\boldsymbol{\nu}} \quad g(\boldsymbol{\lambda}, \boldsymbol{\nu}) \tag{5.26a}$$

$$\text{subject to} \quad \lambda_i \geq 0, \ i = 1, \dots, m. \tag{5.26b}$$

respectively. Now consider to replace each constraint in (5.25b) with $f_i(\mathbf{x}) \leq u_i$ and each constraint in (5.25c) with $h_i(\mathbf{x}) = v_i$. Assume that the objective function in (5.25a) is a measure of cost, and $\mathbf{u} = \{u_i\}_{i=1}^{m}$ and $\mathbf{v} = \{v_i\}_{i=1}^{p}$ are consumed resources. Then, problem

$$\min_{\mathbf{x}} \; f_0(\mathbf{x}) \tag{5.27a}$$

$$\text{subject to } f_i(\mathbf{x}) \leq u_i, \quad i = 1, \ldots, m \tag{5.27b}$$

$$h_i(\mathbf{x}) = v_i, \quad i = 1, \ldots, p \tag{5.27c}$$

is called a *perturbed* problem of (5.25). The corresponding dual problem can be written as

$$\max_{\boldsymbol{\lambda}, \boldsymbol{\nu}} \; g(\boldsymbol{\lambda}, \boldsymbol{\nu}) - \mathbf{u}^T \boldsymbol{\lambda} - \mathbf{v}^T \boldsymbol{\nu} \tag{5.28a}$$

$$\text{subject to } \lambda_i \geq 0, \; i = 1, \ldots, m. \tag{5.28b}$$

The optimal value in the perturbed problem (5.27) is certainly a function of the resources in $\mathbf{u}$ and $\mathbf{v}$, i.e., $p^*(\mathbf{u}, \mathbf{v})$. We are interested in $p^*(\mathbf{u}, \mathbf{v})$, when $\mathbf{u}$ and $\mathbf{v}$ are varying. Under the assumption of zero duality gap, we derive

$$\begin{aligned} p^*(\mathbf{u}, \mathbf{v}) &= g(\boldsymbol{\lambda}^*, \boldsymbol{\nu}^*) - \mathbf{u}^T \boldsymbol{\lambda}^* - \mathbf{v}^T \boldsymbol{\nu}^* \\ &= p^*(0, 0) - \mathbf{u}^T \boldsymbol{\lambda}^* - \mathbf{v}^T \boldsymbol{\nu}^*. \end{aligned} \tag{5.29}$$

In order to evaluate the increase rate of the optimal value $p^*(\mathbf{u}, \mathbf{v})$ in terms of available resources, consider the partial derivatives

$$\frac{\partial p^*(\mathbf{u}, \mathbf{v})}{\partial u_i} = -\lambda_i^* \quad i = 1, \ldots, m \tag{5.30}$$

and

$$\frac{\partial p^*(\mathbf{u}, \mathbf{v})}{\partial v_i} = -\nu_i^* \quad i = 1, \ldots, p. \tag{5.31}$$

Considering these derivations, $\boldsymbol{\lambda}$ and $\boldsymbol{\nu}$ are called *shadow prices* and are derived as

$$\lambda_i^* = -\frac{\partial p^*(\mathbf{u}, \mathbf{v})}{\partial u_i}, \quad \nu_i^* = -\frac{\partial p^*(\mathbf{u}, \mathbf{v})}{\partial v_i}.$$

Recall that $p^*(\mathbf{u}, \mathbf{v})$ is the optimal value of the cost function or equivalently $-p^*(\mathbf{u}, \mathbf{v})$ is the optimal value of the revenue function. Accordingly, the interpretation of the sensitivity analysis is concluded. The larger is the Lagrange multiplier of a constraint in (5.27), the more is the revenue achieved from increasing the resource of this constraint. In other words, it would be more efficient economically to increase the resource of a constraint with the largest Lagrange multiplier.

Example 5 Let $R^*(P_{\max})$ denotes the optimal value of the sum-rate $\sum_{i=1}^{n} r_i$ in problem (5.15) as a function of the maximum allowed transmit power $P_{\max}$. Perform a sensitivity analysis.

Table 5.1 Sensitivity analysis results

P_{max}	0.8	0.9	1	1.1	1.2
$R^*(P_{max})$	0.94	1.044	1.14	1.24	1.34

Solution As derived in the solution of the equivalent problem (5.16) with objective to minimize $-R$, the Lagrange multiplier associated with constraint (5.16b) is $\lambda^* \simeq 1$ for $P_{max} = 1$. By the results of sensitivity analysis, we claim

$$\frac{\partial(-R^*(P_{max}))}{\partial P_{max}} = -1 \tag{5.32}$$

or equivalently

$$\frac{\partial R^*(P_{max})}{\partial P_{max}} = 1. \tag{5.33}$$

To verify this derivation, we show the sum-rate values for a number of maximum allowed transmit powers around $P_{max} = 1$ in Table 5.1. The results verify the derived sensitivity analysis in (5.33), i.e., any variation in P_{max} appears with slope 1 in the sum-rate values. □

5.7 Summary

The process of transforming an optimization problem into a dual domain has been discussed in this chapter. Important concepts of Lagrangian function and dual problem have been introduced. It has been highlighted that, in the case of a minimization problem, the solution of the dual problem provides a useful lower bound for the optimal solution. Based on the results from the dual domain, KKT conditions have been derived, which reduces the problem solution into a set of equality and inequality equations. Furthermore, Lagrangian algorithm as an iterative search method has been proposed, accordingly. Finally, taking advantage of dual domain, a number of application examples in electrical engineering have been introduced.

5.8 Problems

Problem 1 Consider the binary optimization problem

$$\begin{aligned} \min \quad & 2x_1 + x_2 \\ \text{subject to} \quad & x_1 + x_2 \geq 1 \\ & x_1, x_2 \in \{0, 1\} \end{aligned}$$

(a) Determine whether the problem is convex, and find the primal optimal value p^*.
(b) Find the dual function, state the dual problem, and verify that it is a concave problem.
(c) Find the dual optimal value d^* and check out whether the strong duality holds.

Problem 2 Find the dual problem of the following linear program:

$$\min \quad c^T \mathbf{x}$$
$$\text{subject to } \mathbf{Ax} = \mathbf{b}, \mathbf{x} \geq 0.$$

Problem 3 A power station is to supply energy demand E_{tot} of a city during daily hours, index by $\mathcal{T} \triangleq \{t : t = 1, \ldots, 24\}$. The cost of generating x_t kWh energy at time instant t is indicated by $f_t(x_t) = a_t x_t^2 + b_t x_t + c_t$, where $a_t \geq 0$, b_t, c_t are some constants. The objective is to determine $\{x_t\}_{t=1}^{24}$, while minimizing the total cost of power generation and the total daily energy equals E_{tot}.

(a) Formulate the problem as convex optimization problem.
(b) Write down the KKT conditions and form a set of linear equation in the matrix form to find the solution.
(c) Suppose that in addition to the scheduled energy x_t, there is also an unpredicted and time-varying demand over daily hours, modeled as a random variable y. Considering this unknown demand, reformulate the problem in part (a) as a stochastic optimization problem. Is this problem convex?

Problem 4 A radio transmitter is to send data to a receiver through a time-varying communication channel. The channel gain γ is modeled as a random variable varying over a set $\{\gamma_s\}_{s=1}^S$ of states with probabilities $\{\pi_s\}_{s=1}^S$. At each state s with gain γ_s, data transmit rate is $\log_2(1+\gamma_s p_s)$, where p_s is the consumed power by the radio at state s. The objective of the transmitter is to determine transmit powers per states, i.e., $\{p_s\}_{s=1}^S$, while maximizing the average data transmit rate. The average total transmit power is P_T. The problem is formulated as

$$\text{maximize} \sum_{s=1}^{S} \pi_s \log_2(1 + \gamma_s p_s)$$

$$\text{subject to} \sum_{s=1}^{S} \pi_s p_s \leq P_T.$$

(a) Show that the problem is convex.
(b) Write down the KKT conditions for the problem.
(c) Using the KKT conditions, find the optimal solution.

Problem 5 A typical problem in economy is sum-utility maximization. Let two economic resources S_x and S_y to be distributed among N groups of people. The utility of each group i achieved from the share x_i of the resource S_x and the share y_i of the resource S_y is denoted by $\alpha_i \ln(x_i)$ and $\beta_i \ln(y_i)$, respectively, where $\alpha_i > 0$ and $\beta_i > 0$ are weights indicating the priority of group i. The objective is to maximize the total utility subject to the resource constraints. This problem can be written as

$$\begin{aligned} \text{maximize} \; & \sum_{i=1}^{N} \alpha_i \ln(x_i) + \sum_{i=1}^{N} \beta_i \ln(y_i) \\ \text{subject to:} \; & \sum_{i=1}^{N} x_i = S_x \\ & \sum_{i=1}^{N} y_i = S_y \\ & x_i, y_i \geq 0 \quad i = 1, \ldots, N. \end{aligned}$$

(a) Show that the problem is convex.
(b) Write down the KKT conditions and find the optimal solution x_i^* and y_i^* for all i.
(c) Let p^* be the optimal value. Substituting x_i^* and y_i^*, show that $\frac{\partial p^*}{\partial S_x} = \nu_x^*$ and $\frac{\partial p^*}{\partial S_y} = \nu_y^*$, where ν_x^* and ν_x^* are optimal Lagrange multipliers for resource constraints. Therefore, they are rate of increase of the sum-utility value with respect to resource values.
(d) Let $N = 2, \alpha = \beta = [1 \;\; 1]$, and $S_x = 1$, $S_y = 0.5$. Find ν_x^* and ν_y^*. Which resource do you prefer to increase if you are going to increase the sum-utility?

References

1. S. Boyd, L. Vandenberghe, *Convex Optimization* (Cambridge University Press, Cambridge, U.K, 2004)
2. H.W. Kuhn, A.W. Tucker, Nonlinear programming, in *Proceedings of 2nd Berkeley Symposium* (University of California Press, Berkeley, 1951), pp. 481–492
3. Y. Nesterov, A. Nemirovsky, *Interior Point Polynomial Algorithms in Convex Programming*. (SIAM, Philadelphia, 1994)
4. H. Hindi, A tutorial on convex optimization ii: duality and interior point methods, in *Proc. American Control Conference* (2006)
5. A. Goldsmith, *Wireless communications* (Cambridge University Press, New York, 2005)
6. J.F. Kurose, K.W. Ross, *Computer Networking: A Top-Down Approach* (Pearson, New Jersey, 2012)
7. A.S. Tanenbaum, D.J. Wetherall, *Computer Networks* (Prentice Hall, Upper Saddle River, 2010)
8. R. Srikant, L. Ying, *Communication Networks: An Optimization, Control, and Stochastic Networks Perspective* (Cambridge University Press, Cambridge, 2014)
9. M. Chiang, Balancing transport and physical layers in wireless multihop networks: jointly optimal congestion control and power control. Proc. IEEE **23**(1), 104–116 (2005)
10. W. Stanczak, M. Wiczanowski, H. Boche, *Fundamentals of Resource Allocation in Wireless Networks: Theory and Algorithms* (Springer, Berlin, 2008)
11. S. Shakkottai, T.S. Rappaport, P.C. Karlsson, Cross-layer design for wireless networks. IEEE Commun. Mag. **41**(10), 74–80 (2003)
12. M. Chiang, S.H. Low, A.R. Calderbank, J.C. Doyle, Layering as optimization decomposition: a mathematical theory of network architectures. Proc. IEEE **95**(1), 255–312 (2007)

13. F.P. Kelly, A. Maulloo, D. Tan, Rate control for communication networks: shadow prices, proportional fairness and stability. J. Oper. Res. Soc. **49**(3), 237–252 (1998)
14. R. Srikant, *The Mathematics of Internet Congestion Control* (Birkhauser, Basel, 2004)
15. S.H. Low, A duality model of TCP and queue management algorithms. IEEE Trans. Net. **11**(4), 525–536 (2003)
16. W. Yu, R. Lui, Dual methods for nonconvex spectrum optimization of multicarrier systems. IEEE Trans. Commun. **54**(7), 1310–1322 (2006)
17. M. Fathi, H. Taheri, M. Mehrjoo, Cross-layer joint rate control and scheduling for OFDMA wireless mesh networks. IEEE Trans. Veh. Technol. **59**(8), 3933–3941 (2010)

LMI-Based Optimization

6

Abstract

In this chapter, following an introduction on the fundamentals of linear matrix inequalities (LMIs), the application of LMIs to solve convex optimization problems, using numerical examples, is explained. Then, the robust optimization problems are formulated and solved via the LMI-based H_∞ and mixed H_2/H_∞ optimization techniques. In order to solve non-convex optimization problems an iterative LMI is addressed. Finally, the effectiveness of given LMI-based optimization techniques in control synthesis is emphasized and several illustrative examples are given.

Keywords

Linear matrix inequalities (LMI) · Robust optimization · Optimal H_∞ control · Multi-objective optimization · Performance index · Non-convex optimization

6.1 Linear Matrix Inequalities (LMI)

Recently, linear matrix inequalities (LMIs) have emerged as a powerful tool to approach optimization and control problems that appear difficult or impossible to solve in an analytic method. Many optimization problems in dynamic system and control can be formulated as LMI problems. Over the years, efficient algorithms have been developed for solving the LMIs [1].

6.1.1 Lyapunov Inequality

The history of LMIs in the analysis of dynamical systems goes back more than 100 years, when Lyapunov published his work called *Lyapunov theory* in 1890. The basic idea of this theory for stability analysis of linear systems is to search for a positive definite function of the states, called the *Lyapunov function*, whose time derivative is negative definite. A necessary and sufficient condition for the linear system

$$\dot{\mathbf{x}}(t) = \mathbf{A}\mathbf{x}(t) \tag{6.1}$$

M. Fathi, H. Bevrani, *Optimization in Electrical Engineering*,
https://doi.org/10.1007/978-3-030-05309-3_6

where $\mathbf{x} \in \mathbb{R}^n$, $\mathbf{A} \in \mathbb{R}^{n\times n}$, to be stable is the existence of a Lyapunov function $V(\mathbf{x}) = \mathbf{x}^T\mathbf{P}\mathbf{x}$, where $\mathbf{P} \in \mathbb{R}^{n\times n}$ is a symmetric positive definite matrix, such that the time derivative of V is negative for all $\mathbf{x} \neq \mathbf{0}$, i.e.,

$$\begin{aligned}\dot{V}(\mathbf{x}) &= \nabla V^T(\mathbf{x})\dot{\mathbf{x}} = (2\mathbf{P}\mathbf{x})^T\mathbf{A}\mathbf{x} = 2\mathbf{x}^T\mathbf{P}\,\mathbf{A}\mathbf{x} \\ &= \mathbf{x}^T(\mathbf{A}^T\mathbf{P} + \mathbf{P}\mathbf{A})\mathbf{x} < 0 \end{aligned} \tag{6.2}$$

or equivalently

$$\mathbf{A}^T\mathbf{P} + \mathbf{P}\mathbf{A} < 0. \tag{6.3}$$

It is noteworthy that (6.2) is obtained using the scalar values property of $\mathbf{x}^T\mathbf{P}\mathbf{A}\mathbf{x}$, i.e., $\mathbf{x}^T\mathbf{P}\mathbf{A}\mathbf{x} = (\mathbf{x}^T\mathbf{P}\mathbf{A}\mathbf{x})^T = \mathbf{x}^T\mathbf{A}^T\mathbf{P}\mathbf{x}$.

The inequalities (6.3) are called the *Lyapunov inequality* on $\mathbf{P}$, which is a special form of an LMI where $\mathbf{P}$ is variable. Therefore, the first LMI used to analyze stability of a dynamical system was the Lyapunov inequality (6.3), and it is fair to say that Lyapunov is the father of LMI.

Example 1 (Stability Analysis) Evaluate the robust stability of

$$\dot{\mathbf{x}} = \begin{bmatrix} 0 & 1 \\ -\delta & -1 \end{bmatrix}\mathbf{x}, \ \delta \in [0.1 \ \ 1]. \tag{6.4}$$

Solution According to (6.3), we can formulate the stability problem to solve the following LMIs for $\mathbf{P} > 0$:

$$\mathbf{A}_i^T\mathbf{P} + \mathbf{P}\mathbf{A}_i < 0, \quad i = 1, 2$$

This is a convex optimization and can be easily solved in MATLAB environment. □

The LMIs can be used not only for stability analysis but also for control synthesis. For example, it is desired to design state-feedback control law $\mathbf{u}(\mathbf{x}) = \mathbf{K}\mathbf{x}$ for linear system $\dot{\mathbf{x}} = \mathbf{A}\mathbf{x}(t) + \mathbf{B}\mathbf{u}$. The closed-loop system is stable if and only if the LMI (6.3) is satisfied for $\dot{\mathbf{x}}(t) = \mathbf{A}_c\mathbf{x}(t)$, where $\mathbf{A}_c = \mathbf{A} + \mathbf{B}\mathbf{K}$.

$$\mathbf{A}_c^T\mathbf{P} + \mathbf{P}\mathbf{A}_c < 0 \Rightarrow (\mathbf{A} + \mathbf{B}\mathbf{K})^T\mathbf{P} + \mathbf{P}(\mathbf{A} + \mathbf{B}\mathbf{K}) < 0. \tag{6.5}$$

Multiplying both sides by $\mathbf{P}^{-1}$, then

$$\mathbf{A}\mathbf{P}^{-1} + \mathbf{P}^{-1}\mathbf{A}^T + \mathbf{B}\mathbf{K}\mathbf{P}^{-1} + (\mathbf{K}\mathbf{P}^{-1})^T\mathbf{B}^T < 0. \tag{6.6}$$

Changing variables $\mathbf{P}^{-1} = \mathbf{Q}$ and $\mathbf{K}\mathbf{Q} = \mathbf{Y}$, an LMI in terms of $\mathbf{Y}$ and $\mathbf{Q}$ is obtained as

$$\mathbf{A}\mathbf{Q} + \mathbf{Q}\mathbf{A}^T + \mathbf{B}\mathbf{Y} + \mathbf{Y}^T\mathbf{B}^T < 0. \tag{6.7}$$

Now (6.7) is convex and solving it for $\mathbf{Q}$ and $\mathbf{Y}$ yields $\mathbf{K} = \mathbf{Y}\mathbf{Q}^{-1}$ and $\mathbf{P} = \mathbf{Q}^{-1}$.

Example 2 (Controller Synthesis) Design a robust controller for the following uncertain system:

$$\dot{\mathbf{x}} = \begin{bmatrix} 0 & 1 \\ \alpha\delta & 0 \end{bmatrix} \mathbf{x} + \begin{bmatrix} 0 \\ b \end{bmatrix} u, \ \delta \in [0.9 \ 1].$$

Solution

$$\mathbf{A}_1 = \begin{bmatrix} 0 & 1 \\ 0.9\alpha & 0 \end{bmatrix}, \mathbf{A}_2 = \begin{bmatrix} 0 & 1 \\ \alpha & 0 \end{bmatrix},$$

$$\mathbf{B}_1 = \mathbf{B}_2 = \begin{bmatrix} 0 \\ b \end{bmatrix}.$$

Using LMI (6.7) through the MATLAB codes, the robust controller is achieved. □

6.1.2 Standard Presentation

Definition 1 A standard presentation of an LMI is

$$\mathbf{F}(\mathbf{x}) = \mathbf{F}_0 + \sum_{i=1}^{n} x_i \mathbf{F}_i < 0 \tag{6.8}$$

where $\mathbf{F}$ is an *affine function* mapping a finite dimensional vector space Ω to either sets of all $m \times m$ Hermitian $\mathbb{H}$ or symmetric matrices $\mathbb{S}$ ($\mathbf{F} : \Omega \rightarrow \mathbb{H}$ or $\mathbf{F} : \Omega \rightarrow \mathbb{S}$). $\mathbf{x} \in \Omega$ is a vector of n real numbers as the variables. $\mathbf{F}_0, \mathbf{F}_1, \ldots, \mathbf{F}_n$ are real symmetric matrices.

The LMI (6.8) is *feasible* if there exists $\mathbf{x} \in \Omega$ such that $\mathbf{F}(\mathbf{x}) < 0$, otherwise it is said to be *infeasible*. Since all eigenvalues of a real symmetric matrix are real, $\mathbf{F}(\mathbf{x}) < 0$ means $\mathbf{F}(\mathbf{x})$ is negative definite (i.e., $\mathbf{u}^T\mathbf{F}(\mathbf{x})\mathbf{u} < 0$ for all nonzero real vectors $\mathbf{u}$), and all its eigenvalues are negative.

Remark 1 Inequality (6.8) describes a strict LMI. A non-strict LMI is a linear matrix inequality where $<$ in (6.8) is replaced by $\leq$.

Remark 2 An LMI optimization problem is reduced to minimize $\mathbf{c}^T\mathbf{x}$ over all $\mathbf{x} \in \Omega$ that satisfies $\mathbf{F}(\mathbf{x}) < 0$, i.e.,

$$\min_{\mathbf{x}} \ \mathbf{c}^T\mathbf{x} \tag{6.9a}$$

$$\text{subject to } \ \mathbf{F}(\mathbf{x}) < 0. \tag{6.9b}$$

Remark 3 In LMIs, the *matrix variables* can also be used instead of vector variables. This means that we can consider LMIs in the form of $\mathbf{F}(\mathbf{X}) < 0$, where $\mathbf{X} \in \mathbb{R}^{m \times n}$.

A special case with $m = n$ is the Lyapunov inequality $\mathbf{F}(\mathbf{X}) = \mathbf{A}^T\mathbf{X} + \mathbf{X}\mathbf{A} + \mathbf{Q} < 0$, where $\mathbf{A}, \mathbf{Q} \in \mathbb{R}^{n\times n}$ are assumed to be given and $\mathbf{X}$ is the unknown matrix variable of dimension $n \times n$. Note that this defines an LMI only if $\mathbf{Q} \in \mathbb{S}^n$.

Definition 2 A system of LMIs is a finite set of LMIs $\mathbf{F}_1(\mathbf{x}) < 0, \ldots, \mathbf{F}_n(\mathbf{x}) < 0$; with the convex set of all $\mathbf{x}$. These inequalities are satisfied if and only if

$$\mathbf{F}(\mathbf{x}) \triangleq \begin{bmatrix} \mathbf{F}_1(\mathbf{x}) & \mathbf{0} & \ldots & \mathbf{0} \\ \mathbf{0} & \mathbf{F}_2(\mathbf{x}) & \vdots & \mathbf{0} \\ \vdots & \ldots & \ddots & \ldots \\ \mathbf{0} & \mathbf{0} & \vdots & \mathbf{F}_n(\mathbf{x}) \end{bmatrix} < 0. \tag{6.10}$$

Example 3 (Lyapunov Stability) Find an LMI for inequalities $\mathbf{A}^T\mathbf{P} + \mathbf{P}\mathbf{A} < 0$ and $\mathbf{P} > 0$.

Solution According to (6.10), we can present two LMIs to one LMI as follows:

$$\begin{bmatrix} -\mathbf{P} & \mathbf{0} \\ \mathbf{0} & \mathbf{A}^T\mathbf{P} + \mathbf{P}\mathbf{A} \end{bmatrix} < 0.$$ □

Remark 4 The matrix inequalities $\mathbf{F}(\mathbf{x}) > 0$ and $\mathbf{F}(\mathbf{x}) < \mathbf{G}(\mathbf{x})$ with affine functions $\mathbf{F}$ and $\mathbf{G}$ are obtained as special cases of (6.8) as they can be rewritten as the LMIs $-\mathbf{F}(\mathbf{x}) < 0$ and $\mathbf{F}(\mathbf{x}) - \mathbf{G}(\mathbf{x}) < 0$, respectively.

6.1.3 LMI for Convex Optimization

An LMI is a convex constraint. Consequently, optimization problems with convex objective functions can be formulated into LMIs. Therefore, LMIs are a useful tool for solving a wide variety of optimization problems. Linear inequalities, convex quadratic inequalities, matrix norm inequalities, and various constraints from control theory such as Lyapunov and Riccati inequalities can all be written as LMIs. Further, as mentioned, multiple LMIs can always be written as a single LMI of larger dimension. An LMI convex optimization problem is a semi-definite programming (SDP) problem which usually appears in (6.9) or in the form of

$$\min_{\mathbf{x}} \ \mathbf{c}^T\mathbf{x} \tag{6.11a}$$

$$\text{subject to } \ \mathbf{A}(\mathbf{x}) \leq \mathbf{B} \tag{6.11b}$$

where $\mathbf{A}(\mathbf{x}) = x_1\mathbf{A}_1 + \cdots + x_n\mathbf{A}_n$, $\mathbf{x} \in \mathbb{R}^n$, $\mathbf{A}_i$'s and $\mathbf{B}$ are real symmetric matrices.

The constraint $\mathbf{A}(\mathbf{x}) \leq \mathbf{B}$ means $\mathbf{B} - \mathbf{A}(\mathbf{x})$ is positive semi-definite and presents an LMI. For example, Lyapunov theory leads to find a semi-definite matrix $\mathbf{P}$ (i.e., $\mathbf{P} = \mathbf{P}^T$) such that $\mathbf{A}^T\mathbf{P} + \mathbf{P}\mathbf{A} < 0$.

6.2 Equivalent LMIs Representation

6.2.1 Schur Complement

Lemma 1 (Schur Complement) *The LMI*

$$\begin{bmatrix} \mathbf{Q} & \mathbf{S} \\ \mathbf{S}^T & \mathbf{R} \end{bmatrix} < 0 \tag{6.12}$$

is equivalent to the following nonlinear matrix inequalities:

$$\mathbf{Q} < 0,\ \mathbf{R} - \mathbf{S}^T\mathbf{Q}^{-1}\mathbf{S} < 0$$
$$(or\ \mathbf{R} < 0,\ \mathbf{Q} - \mathbf{S}\mathbf{R}^{-1}\mathbf{S}^T < 0). \tag{6.13}$$

Proof Starting from (6.12), the equivalence follows from

$$\begin{bmatrix} \mathbf{I} & \mathbf{0} \\ -\mathbf{S}^T\mathbf{Q}^{-1} & \mathbf{I} \end{bmatrix} \begin{bmatrix} \mathbf{Q} & \mathbf{S} \\ \mathbf{S}^T & \mathbf{R} \end{bmatrix} \begin{bmatrix} \mathbf{I} & -\mathbf{Q}^{-1}\mathbf{S} \\ \mathbf{0} & \mathbf{I} \end{bmatrix}$$
$$= \begin{bmatrix} \mathbf{Q} & \mathbf{0} \\ \mathbf{0} & \mathbf{R} - \mathbf{S}^T\mathbf{Q}^{-1}\mathbf{S} \end{bmatrix} < 0.$$

Similarly, the LMI $\begin{bmatrix} \mathbf{Q} & \mathbf{S} \\ \mathbf{S}^T & \mathbf{R} \end{bmatrix} > 0$ with $\mathbf{Q} = \mathbf{Q}^T > 0$ and $\mathbf{R} = \mathbf{R}^T > 0$ is equivalent to $\mathbf{R} > 0$ and $\mathbf{Q} - \mathbf{S}\mathbf{R}^{-1}\mathbf{S}^T > 0$. □

Example 4 Find an LMI for the constraint $\mathbf{P} > 0,\ \mathbf{c}^T\mathbf{P}^{-1}\mathbf{c} < 1$ where $\mathbf{c} \in \mathbb{R}^n, \mathbf{P} = \mathbf{P}^{-1} \in \mathbb{R}^{n\times n}$.

Solution Using the Schur complement, it is easy to transfer the above two inequalities to $\begin{bmatrix} \mathbf{P} & \mathbf{c} \\ \mathbf{c}^T & 1 \end{bmatrix} > 0$.
□

Example 5 Formulate the following algebraic Riccati inequality to an LMI.

$$\mathbf{A}^T\mathbf{P} + \mathbf{P}\mathbf{A} + \mathbf{P}\mathbf{B}\mathbf{R}^{-1}\mathbf{B}^T\mathbf{P} + \mathbf{Q} < 0$$

where $\mathbf{P}$ is a positive definite symmetric matrix and $\mathbf{Q}$ is a constant symmetric matrix.

Solution The Riccati inequality is quadratic in $\mathbf{P}$ and by applying the Schur complement lemma, one can obtain the following LMI:

$$\begin{bmatrix} -\mathbf{A}^T\mathbf{P} - \mathbf{PA} - \mathbf{Q} & \mathbf{PB} \\ \mathbf{B}^T\mathbf{P} & \mathbf{R} \end{bmatrix} > 0.$$

□

6.2.2 Maximum Singular Value

The *maximum singular value* measures the maximum gain of a multivariable system, which is very useful for quantifying frequency-domain performance and robustness for multivariable systems. The maximum singular value of a matrix $\mathbf{A}$ which depends on $\mathbf{x}$ is denoted by $\bar{\sigma}\mathbf{A}(\mathbf{x})$, which is the square root of the largest eigenvalue of $\mathbf{A}(\mathbf{x})\mathbf{A}^T(\mathbf{x})$ or $\mathbf{A}^T(\mathbf{x})\mathbf{A}(\mathbf{x})$. The inequality $\bar{\sigma}\mathbf{A}(\mathbf{x}) < \gamma$ is a nonlinear convex constraint on $\mathbf{x}$ that may be written as an LMI using the Schur complement lemma:

$$\bar{\sigma}\mathbf{A}(\mathbf{x}) < \gamma \Leftrightarrow \lambda_{\max}\left(\mathbf{A}^T(\mathbf{x})\mathbf{A}(\mathbf{x})\right) < \gamma^2 \Leftrightarrow \frac{1}{\gamma}\mathbf{A}^T(\mathbf{x})\mathbf{A}(\mathbf{x}) - \gamma\mathbf{I} < 0 \Leftrightarrow \begin{bmatrix} \gamma\mathbf{I} & \mathbf{A}(\mathbf{x}) \\ \mathbf{A}^T(\mathbf{x}) & \gamma\mathbf{I} \end{bmatrix} > 0.$$

Example 6 Find an LMI for constraint $\|\mathbf{M}\|_2 < \gamma$, where $\mathbf{M}$ is a given matrix.

Solution $\|\mathbf{M}\|_2 < \gamma \Leftrightarrow \lambda_{\max}\left(\mathbf{M}^T\mathbf{M}\right) < \gamma^2 \Leftrightarrow \mathbf{M}^T\mathbf{M} < \gamma^2\mathbf{I} \Leftrightarrow \gamma^2\mathbf{I} - \mathbf{M}^T\mathbf{M} > 0 \Leftrightarrow \begin{bmatrix} \gamma\mathbf{I} & \mathbf{M} \\ \mathbf{M}^T & \gamma\mathbf{I} \end{bmatrix} > 0.$

□

6.3 LMI-Based Optimization for Robust Control Synthesis

In this section, an LMI-based optimization approach is introduced for robust control synthesis problems. It can be effectively applied for controllers design in a wide range of real-world control systems [2].

6.3.1 Static Output Feedback Control

The static output feedback (SOF) control problem has received so much attention due to its simple structure representation and applicability in the real-world systems. Furthermore, many existing dynamic control synthesis problems can be transferred to an SOF control problem by well-known system augmentation techniques.

Usually, design of a full-order output feedback controller can be reduced to the solution of two convex optimization problems, a state feedback and a Kalman filter. The existence of the SOF controller is shown to be equivalent to the existence of a positive definite matrix simultaneously satisfying two Lyapunov inequalities [3], where the determination of such a matrix leads to solve a non-convex optimization problem [3–6].

Necessary and sufficient conditions for SOF design, as mentioned above, can be obtained in terms of two LMI couples through a bilinear matrix equation [4,5,7]. Particularly, the problem of finding an SOF controller can be restated as a linear algebra problem, which involves two LMIs. For example, an LMI on a positive definite matrix variable $\mathbf{P}$, an LMI on a positive definite matrix variable $\mathbf{Q}$, and a coupling bilinear matrix equation of the form $\mathbf{PQ} = \mathbf{I}$. However, finding such positive definite matrices is a difficult task, since the bilinear matrix equation implies $\mathbf{Q} = \mathbf{P}^{-1}$. Thus, the two LMIs are not convex in $\mathbf{P}$ [4].

6.3.2 H∞-SOF

This section gives a brief overview of $H\infty$-based SOF control design. Consider a linear time invariant system $G(s)$ with the following state-space realization:

$$\dot{\mathbf{x}} = \mathbf{A}\mathbf{x} + \mathbf{B}_1\mathbf{w} + \mathbf{B}_2\mathbf{u} \tag{6.14a}$$

$$\mathbf{z} = \mathbf{C}_1\mathbf{x} + \mathbf{D}_{11}\mathbf{w} + \mathbf{D}_{12}\mathbf{u} \tag{6.14b}$$

$$\mathbf{y} = \mathbf{C}_2\mathbf{x} + \mathbf{D}_{21}\mathbf{w} \tag{6.14c}$$

where $\mathbf{x}$ is the state variable vector, $\mathbf{w}$ is the disturbance and other external input vector, $\mathbf{z}$ is the controlled output vector, and $\mathbf{y}$ is the measured output vector, all in $\mathbb{R}^n$.

The H_∞-based SOF control problem for the linear time invariant system $G(s)$ with the state-space realization of (6.14) is to find a matrix $\mathbf{K}$ (static output feedback law $\mathbf{u} = \mathbf{Ky}$), as shown in Fig. 6.1, such that the resulted closed-loop system is internally stable, and the H_∞ norm from $\mathbf{w}$ to $\mathbf{z}$ is smaller than γ, a specified positive number, i.e.,

$$\|\mathbf{T}_{\mathbf{zw}}(s)\|_\infty < \gamma. \tag{6.15}$$

Under certain assumptions on system matrices, the following theorem can be extendable to the H_∞-SOF control problem.

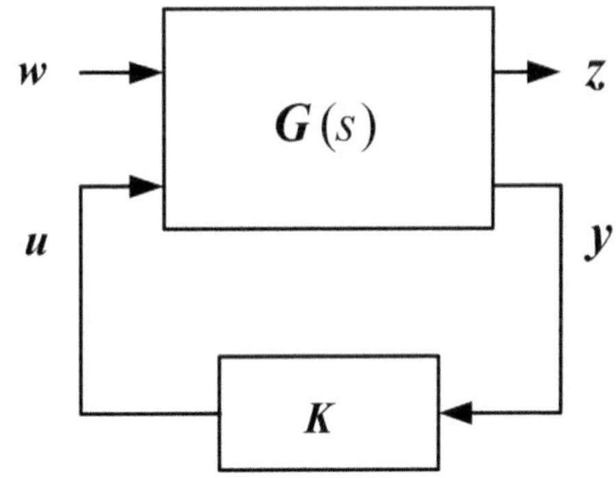

Fig. 6.1 Closed-loop system via H_∞ control

Theorem 1 *It is assumed that* $(\mathbf{A}, \mathbf{B}_2, \mathbf{C}_2)$ *is stabilizable and detectable. The matrix* $\mathbf{K}$ *is a dynamic* H_∞ *controller, if and only if there exists a symmetric matrix* $\mathbf{X} > 0$ *such that*

$$\begin{bmatrix} \mathbf{A}_{\mathrm{cl}}^T\mathbf{X} + \mathbf{X}\mathbf{A}_{\mathrm{cl}} & \mathbf{X}\mathbf{B}_{\mathrm{cl}} & \mathbf{C}_{\mathrm{cl}}^T \\ \mathbf{B}_{\mathrm{cl}}^T\mathbf{X} & -\gamma\mathbf{I} & \mathbf{D}_{\mathrm{cl}}^T \\ \mathbf{C}_{\mathrm{cl}} & \mathbf{D}_{\mathrm{cl}} & -\gamma\mathbf{I} \end{bmatrix} < 0 \tag{6.16}$$

where $\mathbf{A}_{\mathrm{cl}} = \mathbf{A} + \mathbf{B}_2\mathbf{K}\mathbf{C}_2$, $\mathbf{B}_{\mathrm{cl}} = \mathbf{B}_1$, $\mathbf{C}_{\mathrm{cl}} = \mathbf{C}_1 + \mathbf{D}_{12}\mathbf{K}\mathbf{C}_2$, $\mathbf{D}_{\mathrm{cl}} = \mathbf{0}$.

Proof The proof is given in [5] and [8]. □

We can rewrite (6.16) in the following matrix inequality form [9]:

$$\bar{\mathbf{X}}\bar{\mathbf{B}}\mathbf{K}\bar{\mathbf{C}} + (\bar{\mathbf{X}}\bar{\mathbf{B}}\mathbf{K}\bar{\mathbf{C}})^T + \bar{\mathbf{A}}^T\bar{\mathbf{X}} + \bar{\mathbf{X}}\bar{\mathbf{A}} < 0 \tag{6.17}$$

where

$$\bar{\mathbf{A}} = \begin{bmatrix} \mathbf{A} & \mathbf{B}_1 & \mathbf{0} \\ \mathbf{0} & -\gamma\mathbf{I}/2 & \mathbf{0} \\ \mathbf{C}_1 & \mathbf{0} & -\gamma\mathbf{I}/2 \end{bmatrix}, \bar{\mathbf{B}} = \begin{bmatrix} \mathbf{B}_2 \\ \mathbf{0} \\ \mathbf{D}_{12} \end{bmatrix}, \bar{\mathbf{C}} = \begin{bmatrix} \mathbf{C}_2 & \mathbf{0} & \mathbf{0} \end{bmatrix}, \bar{\mathbf{X}} = \begin{bmatrix} \mathbf{X} & \mathbf{0} & \mathbf{0} \\ \mathbf{0} & \mathbf{I} & \mathbf{0} \\ \mathbf{0} & \mathbf{0} & \mathbf{I} \end{bmatrix}. \tag{6.18}$$

Hence, the H_∞-based SOF control problem is reduced to find $\mathbf{X} > 0$ and $\mathbf{K}$ such that matrix inequality (6.17) holds. It is a generalized SOF stabilization problem of the system $(\bar{\mathbf{A}}, \bar{\mathbf{B}}, \bar{\mathbf{C}})$ which can be solved via Theorem 2.

Theorem 2 *The system* $(\mathbf{A}, \mathbf{B}, \mathbf{C})$ *that may also be identified by the following representation:*

$$\dot{\mathbf{x}} = \mathbf{A}\mathbf{x} + \mathbf{B}\mathbf{u} \tag{6.19a}$$

$$\mathbf{y} = \mathbf{C}\mathbf{x} \tag{6.19b}$$

is stabilizable via SOF if and only if there exist $\mathbf{P} > 0$, $\mathbf{X} > 0$ *and* $\mathbf{K}$ *satisfying the following quadratic matrix inequality:*

$$\begin{bmatrix} \mathbf{A}^T\mathbf{X} + \mathbf{X}\mathbf{A} - \mathbf{P}\mathbf{B}\mathbf{B}^T\mathbf{X} - \mathbf{X}\mathbf{B}\mathbf{B}^T\mathbf{P} + \mathbf{P}\mathbf{B}\mathbf{B}^T\mathbf{P} & (\mathbf{B}^T\mathbf{X} + \mathbf{K}\mathbf{C})^T \\ \mathbf{B}^T\mathbf{X} + \mathbf{K}\mathbf{C} & -I \end{bmatrix} < 0. \tag{6.20}$$

Proof According to the Schur complement, the quadratic matrix inequality (6.20) is equivalent to the following matrix inequality:

$$\mathbf{A}^T\mathbf{X} + \mathbf{X}\mathbf{A} - \mathbf{P}\mathbf{B}\mathbf{B}^T\mathbf{X} - \mathbf{X}\mathbf{B}\mathbf{B}^T\mathbf{P} + \mathbf{P}\mathbf{B}\mathbf{B}^T\mathbf{P} + (\mathbf{B}^T\mathbf{X} + \mathbf{K}\mathbf{C})^T(\mathbf{B}^T\mathbf{X} + \mathbf{K}\mathbf{C}) < 0. \tag{6.21}$$

□

Here, the matrices $\mathbf{A}$, $\mathbf{B}$, and $\mathbf{C}$ are constant and have appropriate dimensions. The $\mathbf{X}$ and $\mathbf{P}$ are symmetric and positive definite matrices. For this new inequality notation (6.21), the sufficiency and necessity of theorem are already proven [9].

6.4 An Iterative LMI Algorithm

It is notable that the H_∞-SOF reformulation generally leads to bilinear matrix inequalities (BMI) which are non-convex. This kind of problem is usually solved by an iterative algorithm that may not converge to an optimal solution.

6.4.1 Developed Algorithm

Here, the quadratic matrix inequality (6.20) shows a non-convex optimization problem, and its solution cannot be directly achieved by using general LMI techniques. In order to solve the mentioned H_∞-SOF, an iterative LMI algorithm has been used. The algorithm is mainly based on the idea given in [9]. The key point is to formulate the H_∞ problem via a generalized static output stabilization feedback, such that all eigenvalues of $(\mathbf{A} - \mathbf{BK}_i\mathbf{C})$ shift towards the left half plane through the reduction of a, a real number, to close to feasibility of (6.20). Theorem 2 gives a family of internally stabilizing SOF gains defined as $\mathbf{K}_{\text{sof}}$. A desirable solution can be obtained using the following optimization problem.

Optimization problem: Given an optimal performance index γ (6.15), simply obtained from the application of H_∞ dynamic output feedback control to the control area i (for example, using `hinflmi` function in MATLAB LMI Control Toolbox), determines an admissible SOF law

$$\mathbf{u}_i = \mathbf{K}_i\mathbf{y}_i, \ \mathbf{K}_i \in \mathbf{K}_{\text{sof}} \tag{6.22}$$

such that

$$\|\mathbf{T}_{\mathbf{z}_i\mathbf{w}_i}(s)\|_\infty < \gamma^*, \ \ |\gamma - \gamma^*| < \epsilon \tag{6.23}$$

where ϵ is a small positive number. The performance index γ^* indicates a lower bound such that the closed-loop system is H_∞ stabilizable. The following algorithm gives an iterative LMI (ILMI) solution for the above optimization problem:

Step 1. Set initial values and compute the generalized system $(\bar{\mathbf{A}}_i, \bar{\mathbf{B}}_i, \bar{\mathbf{C}}_i)$ as given in (6.18).

Step 2. Set $i = 1$, $\Delta\gamma = \Delta\gamma_0$ and let $\gamma_i = \gamma_0 > \gamma$. $\Delta\gamma_0$ and γ_0 are positive real numbers.

Step 3. Select $\mathbf{Q} > 0$, and solve $\bar{\mathbf{X}}$ from the following algebraic Riccati equation (ARE)

$$\bar{\mathbf{A}}_i^T\bar{\mathbf{X}} + \bar{\mathbf{X}}\bar{\mathbf{A}}_i - \bar{\mathbf{X}}\bar{\mathbf{B}}_i\bar{\mathbf{B}}_i^T\bar{\mathbf{X}} + \mathbf{Q} = 0. \tag{6.24}$$

then set $\mathbf{P}_1 = \bar{\mathbf{X}}$.

Step 4. *Optimization Problem 1*: Solve the following optimization problem for $\bar{\mathbf{X}}_i$, $\mathbf{K}_i$ and a_i. Minimize a_i subject to the LMI constraints:

$$\begin{bmatrix} \bar{\mathbf{A}}_i^T\bar{\mathbf{X}}_i + \bar{\mathbf{X}}_i\bar{\mathbf{A}}_i - \mathbf{P}_i\bar{\mathbf{B}}_i\bar{\mathbf{B}}_i^T\bar{\mathbf{X}}_i - \bar{\mathbf{X}}_i\bar{\mathbf{B}}_i\bar{\mathbf{B}}_i^T\mathbf{P}_i + \mathbf{P}_i\bar{\mathbf{B}}_i\bar{\mathbf{B}}_i^T\mathbf{P}_i - a_i\bar{\mathbf{X}}_i & (\bar{\mathbf{B}}_i^T\bar{\mathbf{X}}_i + \mathbf{K}_i\bar{\mathbf{C}}_i)^T \\ \bar{\mathbf{B}}_i^T\bar{\mathbf{X}}_i + \mathbf{K}_i\bar{\mathbf{C}} & -\mathbf{I} \end{bmatrix} < 0 \tag{6.25}$$

$$\bar{\mathbf{X}}_i = \bar{\mathbf{X}}_i^T > 0. \tag{6.26}$$

denote a_i^* as the minimized value of a_i.

Step 5. If $a_i^* \leq 0$, go to step 8.

Step 6. For $i > 1$, if $a^*_{i-1} \le 0$, then $\mathbf{K}_{i-1} \in \mathbf{K}_{\text{sof}}$ is an H_∞ controller and $\gamma^* = \gamma_i + \Delta\gamma$ indicates a lower bound such that the above system is H_∞ stabilizable via SOF control, go to step 10.
Step 7. *Optimization Problem 2:* If $i = 1$, solve the optimization problem for $\bar{\mathbf{X}}_i$ and $\mathbf{K}_i$: Minimize $trace(\bar{\mathbf{X}}_i)$ subject to the above LMI constraints (6.25) and (6.26) with $a_i = a^*_i$. Denote $\bar{\mathbf{X}}^*_i$ as the $\bar{\mathbf{X}}_i$ that minimized $trace(\bar{\mathbf{X}}_i)$. Go to step 9.
Step 8. Set $\gamma_i = \gamma_i - \Delta\gamma$, $i = i + 1$. Then go to step 3.
Step 9. If $\|\mathbf{X}_i - \mathbf{P}_i\| < \delta$, ($\delta$ is a prescribed tolerance), go to step 6; else set $i = i+1$ and $\mathbf{P}_i = \bar{\mathbf{X}}^*_{i-1}$, then go to step 4.
Step 10. If the obtained solution ($\mathbf{K}_{i-1}$) satisfies the gain constraint, it is desirable and stop.

The proposed ILMI algorithm shows that if one simply perturbs $\bar{\mathbf{A}}_i$ to $\bar{\mathbf{A}}_i - (a/2)\mathbf{I}$ for some $a > 0$, a solution of the matrix inequality (6.20) can be obtained for the performed generalized plant. That is, there exists a real number ($a > 0$) and a matrix $\mathbf{P} > 0$ to satisfy inequality (6.25). Consequently, the closed-loop system matrix $\bar{\mathbf{A}}_i - \bar{\mathbf{B}}_i\mathbf{K}\bar{\mathbf{C}}_i$ has eigenvalues on the left-hand side of the line $\Re(s) = a$ in the complex s-plane. Based on the idea that all eigenvalues of $\bar{\mathbf{A}}_i - \bar{\mathbf{B}}_i\mathbf{K}\bar{\mathbf{C}}_i$ are shifted progressively towards the left half plane through the reduction of a. The given generalized eigenvalue minimization in the proposed ILMI algorithm guarantees this progressive reduction.

6.4.2 MATLAB Codes

The introduced iterative LMI algorithm can be realized using the MATLAB codes. According to the design approach given in [9], the main program is written as follows.

Main Program

```
%=============================================================
%% Hinf-SOF STABILIZATION ALGORITHM

%% Main function
function [P,K,alfa,i,eigen]=SOF(A,B1,B2,C1,D11,C2,D21,D12,X)


lambda=0.801;                                          %performance index

%% define dimensions
[n,nu]=size(B2);[ny,n]=size(C2);
[m3,n3]=size(B1);[m4,n4]=size(D11);

X=are(A,B2*B2',eye(n));                                %step 3
if min(eig(X))<=0 X=ones(X);end
alfa=1000;alfa1=alfa;K1=zeros(nu,ny);Y=alfa*X;
tt=0;i=1;tol=1e-3;
while i<100000

%% Optimization problem 1 (step 4)
[alfa,K2,X1]=minALFA(A,B1,B2,C1,D11,C2,D21,D12,X,K1,alfa,Y,lambda);

  if alfa<=0                                           %step 5
     P=X1;K=K2;break;

%% if alfa<0 so we have a Hinf SOF stabilizer with performance index of lambda.
 Hence the program stops in step 5.

  elseif alfa>alfa1
        alfa=alfa1;delta=alfa/1000;
```

```
  elseif alfa==alfa1
      delta=alfa/1000;
  elseif alfa1-alfa<tol
      delta=(alfa1-alfa)/4;
  else
      delta=alfa/1000;
  end
  if alfa1-alfa<tol
      tt=tt+1;
  else tt=0;
  end
  if tt>5
      X=X1;K1=K2;
  end

%% Optimization problem 2 (step 7)
 [P,K]=minP(A,B1,B2,C1,D11,C2,D21,D12,X,K1,alfa,lambda);

  while isempty(P)
         alfa=alfa+delta
        [P,K]=minP(A,B1,B2,C1,D11,C2,D21,D12,X,K1,alfa,lambda);
  end

 if norm(X-P)<1e-2
       'convergence,ERROR',break;
  end
  alfa1=alfa,K1=K;X=P;i=i+1
end

%% System (A,B,C)

a=[A B1 zeros(n,m4);zeros(n3,n) -lambda/2*eye(n3) zeros(n3,m4);C1 D11
 -lambda/2*eye(m4)];
b=[B2;zeros(n3,nu);D12];
c=[C2 D21 zeros(ny,m4)];

%%eigenvalues of closed-loop system
eigen=eig(a+b*K*c);
%==============================================================
```

The two optimization problems in the proposed algorithm can be implemented as two separate MATLAB functions (programs). The detailed codes are given below.

Program 1

```
%==============================================================
%% Optimization Problem 1: Minimization of alpha subject to the LMI constraints

function [alfa,Kopt,X1,Y1opt]=...
minALFA(A,B1,B2,C1,D11,C2,D21,D12,X,K0,alfa0,Y0,lambda)

  %% defining dimensions

  [m1,n1]=size(B2);[n2,m1]=size(C2);
  [m3,n3]=size(D11);
  X1=X*B2*B2';X2=X*B2*B2'*X;
```

```
%% defining LMI
setlmis([]);

%%define LMI variables
P=lmivar(1,[m1,1]);
F=lmivar(2,[n1,n2]);
Y1=lmivar(1,[m1,1]);
Y2=lmivar(1,[n3,1]);
Y3=lmivar(1,[m3,1]);

%% LMI #1:
lmiterm([1 1 1 P],A',1,'s');          % A'*P+P*A
lmiterm([1 1 1 0],(X2+X2')/2);        % X*B2*B2'*X
lmiterm([1 1 1 P],X1,-1,'s');         % -X*B2*B2'*P- P*B2*B2'*X
lmiterm([1 2 1 P],B1',1);             % B1'*P
lmiterm([1 2 2 0],-lambda);           % -lambda*I
lmiterm([1 3 1 P],D12*B2',-1);        % -D12*B2'*P
lmiterm([1 3 1 0],C1);                % C1





lmiterm([1 3 2 0],D11);               % D11
lmiterm([1 3 3 0],-lambda);           % -lambda*I
lmiterm([1 3 3 0],-D12*D12');         % -D12*D12'
lmiterm([1 4 1 P],B2',1);             % B2'*P
lmiterm([1 4 1 K],1,C2);              % K*C2
lmiterm([1 4 2 K],1,D21);             % K*D21
lmiterm([1 4 3 0],D12');              % D12'
lmiterm([1 4 4 0],-1);                % -I
lmiterm([-1 1 1 Y1],1,1);             % Y1
lmiterm([-1 2 2 Y2],1,1);             % Y2
lmiterm([-1 3 3 Y3],1,1);             % Y2

%% LMI #2:
lmiterm([-2 1 1 P],1,1);              % P>0

%% LMI #3:
lmiterm([3 1 1 Y1],1,1);              % Y1<alfa*P
lmiterm([3 2 2 Y2],1,1);              % Y2<alfa*I
lmiterm([3 3 3 Y3],1,1);              % Y2<alfa*I
lmiterm([-3 1 1 P],1,1);
lmiterm([-3 2 2 0],1);
lmiterm([-3 3 3 0],1);

%% Eigenvalue minimization
slmi0=getlmis;
options=[1e-3,0, 0,0,1];
xdec=mat2dec(slmi0,X,K0,Y0,alfa0,alfa0);
[tmin,Xopt]=gevp(slmi0,1,options,alfa0,xdec,-1e-6);

%% withdrawing parameters
alfa=tmin;
X1=dec2mat(slmi0,Xopt,P);
Kopt=dec2mat(slmi0,Xopt,K);
Y1opt=dec2mat(slmi0,Xopt,Y1);
Y2opt=dec2mat(slmi0,Xopt,Y2);
%===========================================================
```

Program 2

```
%===========================================================
%% Optimization Problem 2: Minimization of trace(p) subject to the LMI
constraints

function [X1,Kopt]=minP(A,B1,B2,C1,D11,C2,D21,D12,X,K0,alfa,lambda)

  %% defining dimensions
  [m1,n1]=size(B2);[n2,m1]=size(C2);
  X1=X*B2*B2';X2=X1*X;
  %% defining LMI
  setlmis([]);

  %% define LMI variables
  P=lmivar(1,[m1,1]);                   %defining LMI variable
  K=lmivar(2,[n1,n2]);                  %defining LMI variable




  %% LMI #1:
  lmiterm([1 1 1 P],A',1,'s');          % A'*P+P*A
  lmiterm([1 1 1 0],(X2+X2')/2);        % X*B2*B2'*X
lmiterm([1 1 1 P],X1,-1,'s');         % -X*B2*B2'*P-P*B2*B2'*X
  lmiterm([1 1 1 P],-alfa,1);           % -alfa*P
  lmiterm([1 2 1 P],B1',1);             % B1'*P
  lmiterm([1 2 2 0],-alfa);             % -alfa*I
  lmiterm([1 2 2 0],-lambda);            % -lambda*I
  lmiterm([1 3 1 P],D12*B2',-1);        % -D12*B2'*P
  lmiterm([1 3 1 0],C1);                % C1
  lmiterm([1 3 2 0],D11);               % D11
  lmiterm([1 3 3 0],-alfa);             % -alfa*I
  lmiterm([1 3 3 0],-lambda);            % -lambda*I
  lmiterm([1 3 3 0],-D12*D12');         % -D12*D12'*I
  lmiterm([1 4 1 P],B2',1);             % B2'*P
  lmiterm([1 4 1 K],1,C2);              % K*C2
  lmiterm([1 4 2 K],1,D21);             % K*D21
  lmiterm([1 4 3 0],D12');              % D12'
  lmiterm([1 4 4 0],-1);                % -I

  %% LMI #2:
  lmiterm([-2 1 1 P],1,1);              % P>0

  %% solve the minimization problem
  slmi1=getlmis;
  n=decnbr(slmi1);c=zeros(n,1);
  for j=1:n
       pj=defcx(slmi1,j,P); c(j)=trace(pj);
  end,
  options=[1e-5,0,0,0,1];
  xdec=mat2dec(slmi1,X,K0);
[copt,Xopt]=mincx(slmi1,c,options,xdec,1e-10);
  if isempty(Xopt),X1=[];Kopt=[];
  else

  %% withdrawing parameters
     X1=dec2mat(slmi1,Xopt,P);
     Kopt=dec2mat(slmi1,Xopt,K);
  end
%===========================================================
```

6.4.3 Numerical Examples

Example 7 Design an H_∞-based SOF controller for the following dynamic system:

$$\dot{x} = -x + \begin{bmatrix}1 & 0\end{bmatrix} w + u \tag{6.27a}$$

$$z = \begin{bmatrix}1 & 0\end{bmatrix} x + \begin{bmatrix}0 \\ 1\end{bmatrix} u \tag{6.27b}$$

$$y = x + \begin{bmatrix}0 & 1\end{bmatrix} w. \tag{6.27c}$$

Solution Setting the performance index at 0.9 gives $\alpha = -0.0164$ and $K = -0.3576$ after two iterations. For performance index of 0.802, after seven iterations, we can obtain $\alpha = -5.3148e^{-6}$ and $K = -0.5499$. $\gamma = 0.802$ is the solution performance index, because by setting it at 0.801, we cannot get a feasible answer from the algorithm. □

```
%============================================================
A=-1;
B1=[1 0];
B2=1;
C1=B1';
C2=1;
D11=zeros(2);
D12=[0;1];
D21=[0 1];

% Raining the Hinf SOF
[P,K,alfa,iterations,eigenvalues]=SOF(A,B1,B2,C1,D11,C2,D21,D12);
P,K,alfa,iterations,eigenvalues

P =
    1.2355
K =
   -0.5499
alfa =
  -5.3148e-06
iterations =
     7
eigenvalues =
   -0.4010
   -0.4010
   -1.5499
   -0.4010
   -0.4010
%============================================================
```

Example 8 Design an H_∞-based SOF controller for the following dynamic system.

$$\dot{\mathbf{x}} = \mathbf{A}\mathbf{x} + \mathbf{B}_1\mathbf{w} + \mathbf{B}_2\mathbf{u} \tag{6.28a}$$

$$\mathbf{z} = \mathbf{C}_1\mathbf{x} + \mathbf{D}_{12}\mathbf{u} \tag{6.28b}$$

$$\mathbf{y} = \mathbf{C}_2\mathbf{x} \tag{6.28c}$$

where

$$\mathbf{A} = \begin{bmatrix} 0 & 0 & 1.132 & 0 & -1 \\ 0 & -0.0538 & -0.1712 & 0 & 0.705 \\ 0 & 0 & 0 & 1 & 0 \\ 0 & 0.0485 & 0 & -0.8556 & -1.013 \\ 0 & -0.2909 & 0 & -1.0532 & -0.6859 \end{bmatrix}, \; \mathbf{B}_1 = \begin{bmatrix} 0.03593 & 0 & 0.01672 \\ 0 & 0.00989 & 0 \\ 0 & 0.07548 & 0 \\ 0 & 0 & 0.05635 \\ 0.00145 & 0 & 0.06743 \end{bmatrix},$$

$$\mathbf{B}_2 = \begin{bmatrix} 0 & 0 & 0 \\ -0.12 & 1 & 0 \\ 0 & 0 & 0 \\ 4.419 & 0 & -1.665 \\ 1.575 & 0 & -0.0732 \end{bmatrix}, \mathbf{C}_1 = \begin{bmatrix} 0 & d & 0 & 0 & 0 \\ 0 & 0 & d & 0 & 0 \\ 0 & 0 & 0 & 0 & 0 \\ 0 & 0 & 0 & 0 & 0 \\ 0 & 0 & 0 & 0 & 0 \end{bmatrix}, \mathbf{C}_2 = \begin{bmatrix} 1 & 0 & 0 & 0 & 0 \\ 0 & 1 & 0 & 0 & 0 \\ 0 & 0 & 1 & 0 & 0 \end{bmatrix}, \mathbf{D}_{12} = \begin{bmatrix} 0 & 0 & 0 \\ 0 & 0 & 0 \\ d & 0 & 0 \\ 0 & d & 0 \\ 0 & 0 & d \end{bmatrix},$$

$d = 0.707106781186548.$

Solution This example shows the dynamic model of an aircraft [10]. The robust performance index is determined as $\gamma = 0.12$, and after fifteen iterations the solution is achieved. □

```
%=============================================================
A=[0,0,1.132,0,−1;0,−0.0538,−0.1712,0,0.0705;
     0,0,0,1,0;0,0.0485,0,−0.8556,−1.013;0,−0.2909,0,1.0532,−0.6859];
 B2=[0,0,0;−0.12,1,0;0,0,0;4.419,0,−1.665;1.575,0,−0.0732];
 B1=[0.03593,0,0.01672;0,0.00989,0;0,−0.07548,0;0,0,0.05635;0.00145,0,0.06743];
 C2=[1,0,0,0,0;0,1,0,0,0;0,0,1,0,0];
 C1=[0,0.707106781186548,0,0,0;0,0,0.707106781186548,0,0;...
 0,0,0,0,0;0,0,0,0,0;0,0,0,0,0];
 D11=[0,0,0;0,0,0;0,0,0;0,0,0;0,0,0];
 D12=[0,0,0;0,0,0;0.707106781186548,0,0;0,0.707106781186548,0;...
 0,0,0.707106781186548];
 D21=[0,0,0;0,0,0;0,0,0];

 % Raining the Hinf SOF
 [P,K, alfa , iterations , eigenvalues]=SOF(A,B1,B2,C1,D11,C2,D21,D12);
 P,K, alfa , iterations , eigenvalues

P =
    0.1671     0.2911     0.9099     0.2009    −0.4038
    0.2911     6.9825     1.8165    −0.0895    −0.3185
    0.9099     1.8165    11.6034     2.7984    −3.6254
    0.2009    −0.0895     2.7984     1.6434    −1.5504
   −0.4038    −0.3185    −3.6254    −1.5504     3.4396
```

```
K =
    -0.0565    -0.5282    -1.0339
    -0.0531    -1.1752    -0.3750
     0.0623    -0.1121     0.8632
alfa =
   -1.4858e-04
iterations =
    15
eigenvalues =
  -0.0600 + 0.0000i
  -0.0600 + 0.0000i
  -0.0600 + 0.0000i
  -0.0600 + 0.0000i
  -0.0600 + 0.0000i
  -0.6606 + 2.5989i
  -0.6606 - 2.5989i
  -0.1873 + 0.0770i
  -0.1873 - 0.0770i
  -1.0113 + 0.0000i
  -0.0600 + 0.0000i
  -0.0600 + 0.0000i
  -0.0600 + 0.0000i
%===========================================================
```

6.5 LMI-Based Robust Multi-Objective Optimization

In many real-world control problems, it is desirable to follow several objectives such as stability, disturbance attenuation, reference tracking, and considering the practical constraints, simultaneously. Pure H_∞ synthesis cannot adequately capture all design specifications. For instance, H_∞ synthesis mainly enforces closed-loop stability and meets some constraints and limitations, while noise attenuation or regulation against random disturbances is more naturally expressed in linear quadratic Gaussian (LQG) terms (H_2 synthesis). The mixed H_2/H_∞ control synthesis gives a powerful multi-objective control design addressed by the LMI techniques.

6.5.1 Optimal Mixed H_2/H_∞ Control Design

A general synthesis control scheme using a mixed H_2/H_∞ control technique [2] is shown in Fig. 6.2. $G(s)$ is a linear time invariant system with the following state-space realization:

$$\dot{\mathbf{x}} = \mathbf{A}\mathbf{x} + \mathbf{B}_1\mathbf{w} + \mathbf{B}_2\mathbf{u} \tag{6.29a}$$

$$\mathbf{z}_\infty = \mathbf{C}_\infty\mathbf{x} + \mathbf{D}_{\infty 1}\mathbf{w} + \mathbf{D}_{\infty 2}\mathbf{u} \tag{6.29b}$$

$$\mathbf{z}_2 = \mathbf{C}_2\mathbf{x} + \mathbf{D}_{21}\mathbf{w} + \mathbf{D}_{22}\mathbf{u} \tag{6.29c}$$

$$\mathbf{y} = \mathbf{C}_y\mathbf{x} + \mathbf{D}_{y1}\mathbf{w} \tag{6.29d}$$

where $\mathbf{x}$ is the state variable vector, $\mathbf{w}$ is the disturbance and other external input vector, and $\mathbf{y}$ is the measured output vector. The output channel $\mathbf{z}_2$ is associated with the LQG aspects (H_2 performance), while the output channel $\mathbf{z}_\infty$ is associated with the H_∞ performance. Let $T_\infty(s)$ and $T_2(s)$ be the

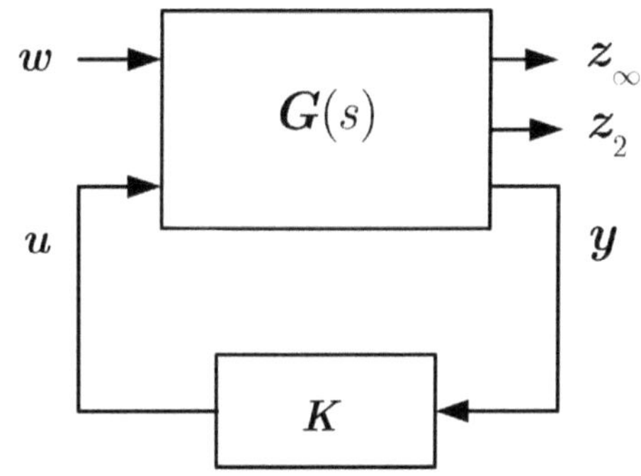

Fig. 6.2 Closed-loop system via mixed H_2/H_∞ control

transfer functions from $\mathbf{w}$ to $\mathbf{z}_\infty$ and $\mathbf{z}_2$, respectively, and consider the following state space realization for the closed-loop system:

$$\dot{\mathbf{x}}_{\text{cl}} = \mathbf{A}_{\text{cl}}\mathbf{x}_{\text{cl}} + \mathbf{B}_{\text{cl}}\mathbf{w} \tag{6.30a}$$

$$\mathbf{z}_\infty = \mathbf{C}_{\text{cl1}}\mathbf{x}_{\text{cl}} + \mathbf{D}_{\text{cl1}}\mathbf{w} \tag{6.30b}$$

$$\mathbf{z}_2 = \mathbf{C}_{\text{cl2}}\mathbf{x}_{\text{cl}} + \mathbf{D}_{\text{cl2}}\mathbf{w} \tag{6.30c}$$

$$\mathbf{y} = \mathbf{C}_{\text{cl}}\mathbf{x}_{\text{cl}} + \mathbf{D}_{\text{cl}}\mathbf{w} \tag{6.30d}$$

Theorems 3 and 4 express the design objectives in terms of LMIs. Interested readers can find more details and proofs in [11–13].

Theorem 3 (**H_∞ Performance**) *The closed-loop RMS gain for $T_\infty(s)$ does not exceed γ_∞ if and only if there exists a symmetric matrix $\mathbf{X}_\infty > 0$ such that*

$$\begin{bmatrix} \mathbf{A}_{\text{cl}}\mathbf{X}_\infty + \mathbf{X}_\infty\mathbf{A}_{\text{cl}}^T & \mathbf{B}_{\text{cl}} & \mathbf{X}_\infty\mathbf{C}_{\text{cl1}}^T \\ \mathbf{B}_{\text{cl1}}^T & -\mathbf{I} & \mathbf{D}_{\text{cl1}}^T \\ \mathbf{C}_{\text{cl1}}\mathbf{X}_\infty & \mathbf{D}_{\text{cl1}} & -\gamma_\infty^2\mathbf{I} \end{bmatrix} < 0. \tag{6.31}$$

Theorem 4 (**H_2 Performance**) *The H_2 norm of $T_2(s)$ does not exceed γ_2 if and only if $\mathbf{D}_{\text{cl2}} =$ $\mathbf{0}$, and there exist two symmetric matrices $\mathbf{X}_2$ and $\mathbf{Q}$ such that*

$$\begin{bmatrix} \mathbf{A}_{\text{cl}}\mathbf{X}_2 + \mathbf{X}_2\mathbf{A}_{\text{cl}}^T & \mathbf{B}_{\text{cl}} \\ \mathbf{B}_{\text{cl1}}^T & -\mathbf{I} \end{bmatrix} < 0, \begin{bmatrix} \mathbf{Q} & \mathbf{C}_{\text{cl2}}\mathbf{X}_2 \\ \mathbf{X}_2\mathbf{C}_{\text{cl2}}^T & \mathbf{X}_2 \end{bmatrix} > 0, \mathit{Trace}(\mathbf{Q}) < \gamma_2^2. \tag{6.32}$$

The mixed H_2/H_∞ control design method uses both theorems and gives us an output feedback controller $K(s)$ that minimizes the following trade-off criterion:

$$k_1\|T_\infty(s)\|_\infty^2 + k_2\|T_2(s)\|_2^2, \quad (k_1 \geq 0, k_2 \geq 0). \tag{6.33}$$

An efficient algorithm to solve this problem is available in function `hinfmix` of the LMI control toolbox for MATLAB [14].

Theorem 5 gives the necessary and sufficient condition for the existence of the H_∞-based SOF controller to meet the following performance criteria:

$$\|T_\infty(s)\|_\infty < \gamma_\infty. \tag{6.34}$$

Similarly, Theorem 4 gives the necessary and sufficient condition for the existence of the H_2-based SOF controller to meet

$$\|T_2(s)\|_2 < \gamma_2 \tag{6.35}$$

where γ_2 is the H_2 optimal performance index, which demonstrates the minimum upper bound of H_2 norm and specifies the disturbance attention level.

Theorem 5 *For fixed* $(\mathbf{A}, \mathbf{B}_1, \mathbf{B}_2, \mathbf{C}_y, \mathbf{K})$, *there exist a positive definite matrix* $\mathbf{X}$ *which solves inequality*

$$\begin{aligned} &(\mathbf{A} + \mathbf{B}_2\mathbf{K}\mathbf{C}_y)\mathbf{X} + \mathbf{X}(\mathbf{A} + \mathbf{B}_2\mathbf{K}\mathbf{C}_y)^T + \mathbf{B}_1\mathbf{B}_1^T < 0 \\ &\mathbf{X} > L_c \end{aligned} \tag{6.36}$$

to satisfy (6.35), if and only if the following inequality has a positive definite matrix solution [15]:

$$\mathbf{A}\mathbf{X} + \mathbf{X}\mathbf{A}^T - \mathbf{X}\mathbf{C}_y^T\mathbf{C}_y\mathbf{X} + (\mathbf{B}_2\mathbf{K} + \mathbf{X}\mathbf{C}_y^T)(\mathbf{B}_2\mathbf{K} + \mathbf{X}\mathbf{C}_y^T)^T + \mathbf{B}_1\mathbf{B}_1^T < 0 \tag{6.37}$$

where L_C *in (6.36) denotes the controllability Gramian matrix of the pair* $(\mathbf{A}_c, \mathbf{B}_{1c})$ *and can be related to the* H_2 *norm presented in (6.35) as follows:*

$$\|T_2(s)\|_2^2 = \text{trace}(\mathbf{C}_{2c}L_C\mathbf{C}_{2c}^T) \tag{6.38}$$

Notice that the condition that $\mathbf{A} + \mathbf{B}_2\mathbf{K}\mathbf{C}_y$ is Hurwitz is implied by inequality (6.36). Thus if

$$\text{trace}(\mathbf{C}_{2c}\mathbf{X}\mathbf{C}_{2c}^T) < \gamma_2^2 \tag{6.39}$$

the requirement is satisfied.

6.5.2 An Iterative LMI Algorithm

In the proposed control strategy, to design mixed H_2/H_∞ SOF controller and to solve the yielding non-convex optimization problem, which cannot be directly achieved by using LMI techniques, an ILMI algorithm is developed.

The optimization problem given in (6.33) defines a robust performance synthesis problem, where the H_2 norm is chosen as the performance measure. Recently, several LMI-based methods are proposed to obtain the suboptimal solution for H_2, H_∞ and/or H_2/H_∞ SOF control problems. It is noteworthy that using Theorem 5 is difficult to directly achieve a solution for (6.37) by the general LMI. Here, an ILMI algorithm is introduced to get a suboptimal solution for the above optimization problem. Specifically, the proposed algorithm formulates the H_2/H_∞ SOF control through a general SOF stabilization problem. The proposed algorithm searches the desired suboptimal H_2/H_∞ SOF controller $\mathbf{K}_i$ within a family of H_2 stabilizing controllers $\mathbf{K}_{sof}$ such that

$$|\gamma_2^* - \gamma_2| < \epsilon, \ \gamma_\infty = \|T_\infty(s)\|_\infty < 1 \tag{6.40}$$

where ϵ is a small real positive number, γ_2^* is H_2 performance corresponding to H_2/H_∞ SOF controller $\mathbf{K}_i$, and γ_2 is optimal H_2 performance index that can result from the application of standard H_2/H_∞ dynamic output feedback control.

In the proposed strategy, first the stability domain of SOF parameters space is specified, which guarantees the stability of closed-loop system. In the second step, the subset of the stability domain in the SOF parameter space is specified to minimize the H_2 tracking performance. Finally, the design problem becomes the point with the closest H_2 performance index to an optimal one which meets the H_∞ constraint. The main effort is to formulate the H_2/H_∞ problem via the generalized static output stabilization feedback lemma such that all eigenvalues of $(\mathbf{A} - \mathbf{BKC})$ shift towards the left half plane through the reduction of a_i, a real number, to close to the feasibility of (6.33). The proposed algorithm is schematically described in Fig. 6.3.

6.6 Summary

In this chapter, the definition and characteristics of LMI were briefly introduced. It was shown that many optimization problems in real-world engineering can be formulated as LMI problems. A useful LMI transformation was discussed and the application of LMIs to solve convex optimization problems, using numerical examples, was explained. Finally, the robust optimization problems were formulated and solved via the LMI-based H_∞ and mixed H_2/H_∞ control methodologies. In order to solve non-convex optimization problems iterative LMIs were developed. The effectiveness of the given LMI-based optimization techniques in control synthesis was emphasized and several illustrative examples with MATLAB codes were given.

6.7 Problems

Problem 1 Determine strict and non-strict LMIs:

(a) $\mathbf{P} > 0$.
(b) $\mathbf{PA} + \mathbf{A}^T\mathbf{P} < 0$.
(c) $\begin{bmatrix} -\mathbf{A}^T\mathbf{P} - \mathbf{PA} - \mathbf{C}^T\mathbf{C} & -\mathbf{PB} \\ -\mathbf{B}^T\mathbf{P} & \gamma\mathbf{I} \end{bmatrix} \geq 0$.

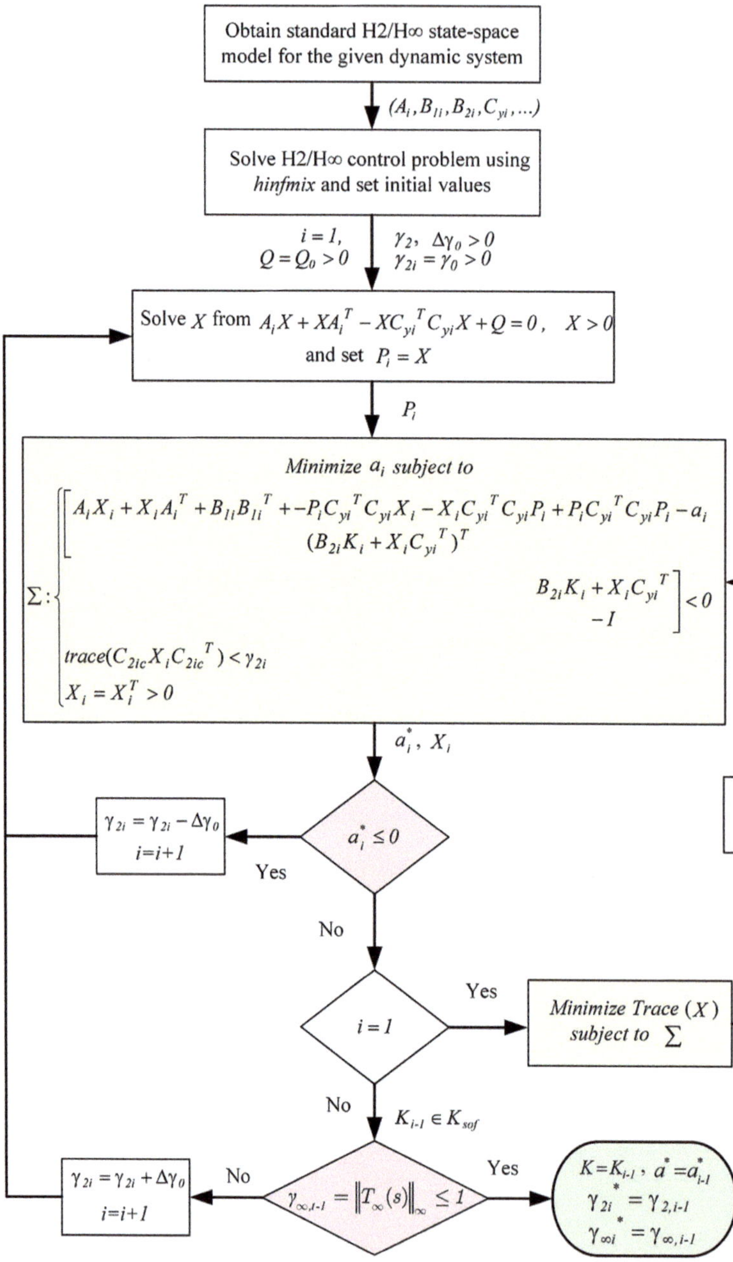

Fig. 6.3 Iterative LMI algorithm for mixed H_2/H_∞ SOF control [2]

Problem 2 Generate MATLAB codes to solve Examples 1 and 2.

Problem 3 Find an equivalent LMI for the following ellipsoid constraints:

$$\mathbf{P} = \mathbf{P}^T > 0, (\mathbf{x} - \mathbf{x}_0)^T \mathbf{P}^{-1} (\mathbf{x} - \mathbf{x}_0) < 1; \mathbf{x} \in \mathbb{R}^n, \mathbf{P} \in \mathbb{R}^{n \times n}$$

Problem 4 Find the Schur complement form of the LMI $(\mathbf{C} - \lambda\mathbf{F})(\lambda\mathbf{E} - \mathbf{A})^{-1}\mathbf{B} + \mathbf{D} < 0$.

Problem 5 Find an equivalent LMI for the strict inequality $\mathbf{E}_0(\mathbf{x}) < \tau\mathbf{E}_1(\mathbf{x})$, for all $\mathbf{x} \in \mathbb{R}^n$ and $\tau \geq 0$, where $\mathbf{E}_0(\mathbf{x}) = \mathbf{x}^T\mathbf{F}_0\mathbf{x} + 2\mathbf{g}_0^T\mathbf{x} + \mathbf{h}_0$ and $\mathbf{E}_1(\mathbf{x}) = \mathbf{x}^T\mathbf{F}_1\mathbf{x} + 2\mathbf{g}_1^T\mathbf{x} + \mathbf{h}_1$.

Problem 6 Show that the inequalities

$$\mathbf{P} > 0, \lambda \geq 0, \begin{bmatrix} \mathbf{A}^T\mathbf{P}+\mathbf{P}\mathbf{A}+\lambda\mathbf{C}^T\mathbf{C} & \mathbf{P}\mathbf{B}+\lambda\mathbf{C}^T\mathbf{D} \\ \mathbf{B}^T\mathbf{P}+\lambda\mathbf{D}^T\mathbf{C} & -\lambda(\mathbf{I}-\mathbf{D}^T\mathbf{D}) \end{bmatrix} < 0.$$

are equivalent to

$$\mathbf{Q} > 0, \mu \geq 0, \begin{bmatrix} \mathbf{Q}\mathbf{A}^T + \mathbf{A}\mathbf{Q} + \mu\mathbf{B}\mathbf{B}^T & \mathbf{Q}\mathbf{C}^T + \mu\mathbf{B}\mathbf{D}^T \\ \mathbf{C}\mathbf{Q} + \mu\mathbf{D}\mathbf{B}^T & \mu(\mathbf{D}\mathbf{D}^T - \mathbf{I}) \end{bmatrix} < 0.$$

Problem 7 Find an equivalent LMI for the following linear differential inclusions:

$$\mathbf{E}_0(\mathbf{x}) = \mathbf{x}^T\mathbf{P}(\mathbf{A}\mathbf{x} + \mathbf{B}\mathbf{P}) + (\mathbf{A}\mathbf{x} + \mathbf{B}\mathbf{P})^T\mathbf{P}\mathbf{x} < 0$$

$$\mathbf{E}_1(\mathbf{x}, \mathbf{P}) = \mathbf{P}^T\mathbf{P} - (\mathbf{C}\mathbf{x} + \mathbf{D}\mathbf{P})^T(\mathbf{C}\mathbf{x} + \mathbf{D}\mathbf{P}) \leq 0, \mathbf{x}, \mathbf{P} \neq \mathbf{0}$$

$$\mathbf{P} > 0, \lambda \geq 0, \mathbf{E}_0(\mathbf{x}) < \lambda\mathbf{E}_1(\mathbf{x}, \mathbf{P}), \forall\mathbf{x}, \mathbf{P}.$$

Problem 8 Design an H_∞-based SOF controller for the following dynamic system, while the robust performance index is fixed at $\gamma = 0.96$.

$$\dot{\mathbf{x}} = \begin{bmatrix} 0 & 1 \\ 1 & 0 \end{bmatrix} \mathbf{x} + \begin{bmatrix} 1 \\ 0 \end{bmatrix} u$$

$$y = \begin{bmatrix} 1 & 15 \end{bmatrix} \mathbf{x}.$$

Problem 9 Design an H_∞-based SOF controller for the following dynamic system, while the robust performance index is fixed at $\gamma = 0.7$.

$$\dot{\mathbf{x}} = \mathbf{A}\mathbf{x} + \mathbf{B}_2\mathbf{u}$$

$$\mathbf{z} = \mathbf{D}_{12}\mathbf{u}$$

$$\mathbf{y} = \mathbf{C}_2\mathbf{x} + \mathbf{D}_{21}\mathbf{w}$$

where $\mathbf{A} = \begin{bmatrix} -0.0366 & 0.0271 & 0.0188 & -0.4555 \\ 0.0482 & -1.01 & 0.0024 & -4.0208 \\ 0.1002 & 0.3681 & -0.7070 & 1.42 \\ 0 & 0 & 1 & 0 \end{bmatrix}$, $\mathbf{B}_2 = \begin{bmatrix} 0.4422 & 0.1761 \\ 3.5446 & -7.5922 \\ -5.52 & 4.49 \\ 0 & 0 \end{bmatrix}$,

$\mathbf{C}_2 = \begin{bmatrix} 0 & 1 & 0 & 0 \end{bmatrix}, \mathbf{D}_{12} = \begin{bmatrix} 1 & 0 \end{bmatrix}, \mathbf{D}_{21} = 1.$

Problem 10 Generate MATLAB codes for the given LMI-based multi-objective control design algorithm in Fig. 6.3.

References

1. S. Boyd, L.E. Ghaoui, E. Feron, V. Balakrishnan, *Linear Matrix Inequalities in Systems and Control Theory* (Philadelphia, SIAM Books, 1994)
2. H. Bevrani, *Robust Power System Frequency Control* (Basel, Springer, 2014)
3. T. Iwasaki, R.E. Skelton, J.C. Geromel, Linear quadratic suboptimal control with static output feedback. Syst. Control Lett. **23**, 421–430 (1994)
4. F. Leibfritz, *Static Output Feedback Design Problems*, PhD Dissertation, Trier University, 1998
5. P. Gahinet, P. Apkarian, A linear matrix inequality approach to H_∞ control. Int. J. Robust Nonlinear Control. **4**, 421–448 (1994)
6. J.C. Geromel, C.C. Souza, R.E. Skeltox, Static output feedback controllers: stability and convexity. IEEE Trans. Autom. Control. **42**, 988–992 (1997)
7. T. Iwasaki, R.E. Skelton, All controllers for the general H_∞ control problem: LMI existence conditions and state space formulas. Automatica. **30**, 1307–1317 (1994)
8. R.E. Skelton, J. Stoustrup, T. Iwasaki, The H_∞ control problem using static output feedback. Int. J. Robust Nonlinear Control. **4**, 449–455 (1994)
9. Y.Y. Cao, Y.X. Sun, W.J. Mao, Static output feedback stabilization: an ILMI approach. Automatica **34**, 1641–1645 (1998)

10. Y.S. Hung, A.G.J. MacFarlane, Multivariable feedback: a quasi-classical approach, in: *Lecture Notes in Control and Information Sciences* (Springer, New York, 1982)
11. P.P. Khargonekar, M.A. Rotea, Mixed H2/H_∞ control: a convex optimization approach. IEEE Trans. Autom. Control. **39**, 824–837 (1991)
12. C.W. Scherer, Multiobjective H2/H_∞ control. IEEE Trans. Autom. Control. **40**, 1054–1062 (1995)
13. C.W. Scherer, P. Gahinet, M. Chilali, Multiobjective output-feedback control via LMI optimization. IEEE Trans. Autom. Control. **42**, 896–911 (1997)
14. P. Gahinet, A. Nemirovski, A.J. Laub, M. Chilali. *LMI Control Toolbox* (The MathWorks, Natick, 1995)
15. F. Zheng, Q.G. Wang, H.T. Lee, On the design of multivariable PID controllers via LMI approach. Automatica **38**, 517–526 (2002)

7 Artificial Intelligence and Evolutionary Algorithms-Based Optimization

Abstract

Recently, applications of artificial intelligence (AI) techniques and evolutionary algorithms (EA) have received increasing attention in engineering optimization problems. Numerous research works indicate the applicability of these approaches on the optimization issues. While many of these approaches are still under investigation, due to significant advances in metering, computing, and communication technologies, there already exist a number of practical optimization across the engineering world. The present chapter explains the power of AI methodologies and EAs in engineering optimization problems. The basics of some AI and EA techniques are described, and the state of the art of these optimization methodologies in engineering applications is presented. In particular, the application structures of artificial neural networks, particle swarm optimization, and genetic algorithm in engineering optimization problems are explained. Given optimization approaches are supplemented by several examples.

Keywords

Artificial intelligence · Evolutionary algorithm · Artificial neural network · Particle swarm optimization · Genetic algorithm · Search methods · Gradient-based optimization · Multi-objective optimization · Pareto-optimal solution · Fitness function · Learning algorithm · Backpropagation

7.1 Introduction

The fact that most optimization problems, where modeled accurately, are of non-convex and sometimes discrete nature has encouraged many researchers to develop new optimization methodologies to overcome such difficulties. Concerning the limitations for the application of well-known convex optimization approaches, discussed in the previous chapters, for real-world engineering systems attract increasing interest on the applications of AI and EAs such as knowledge-based expert systems, artificial neural networks (ANNs), neuro-fuzzy systems, genetic algorithms (GA), searching-based evolution algorithms, and other intelligent technologies. This is mostly due to real systems uncertainties and intermittency in one side, and the simplicity, smartness, and flexibility of artificial intelligent approaches, in the other side.

M. Fathi, H. Bevrani, *Optimization in Electrical Engineering*,
https://doi.org/10.1007/978-3-030-05309-3_7

Over the last few decades, many optimization applications have been proposed in the literature to demonstrate the advantages of AI and EAs over the conventional methodologies. Recently, following the advent of modern intelligent methods, such as ANNs, fuzzy logic, multi-agent systems (MASs), GAs, expert systems, simulated annealing, Tabu search, particle swarm optimization (PSO), ant colony optimization, and hybrid intelligent techniques, some new potentials and powerful solutions for solving the real-world engineering optimization problems have arisen. A number of EAs- and AI-based optimization approaches have been currently utilized in the electric power grids. Some application areas of intelligent technologies in Japanese power system utilities are reported in [1].

In the EAs and AI methodologies, an optimization problem in system planning, operation, control, or management usually reduces to optimize (finding minimum or maximum) a given cost function or fitness function. These approaches offer many benefits in the area of complex and non-convex optimization problems, particularly for the systems operating over an uncertain operating range. The human and nature ability to optimize complex organisms operation has encouraged researchers to pattern optimization methodologies on human/nature responses, fuzzy behaviors, and neural network systems. Since all of developed AI techniques are usually dependent on knowledge extracted from environment and available data, the *knowledge management* plays pivotal role in intelligent-based optimization procedures.

In the present chapter, due to the lake of space, among the existing AI techniques and EAs, we will focus on the ANNs-, PSO-, and GA-based optimization methodologies. Indeed, an ANN consists of a finite number of interconnected neurons and acts as a massively parallel distributed processor, inspired from biological neural networks, which can store experimental knowledge and makes it available for use. The ANNs are parallel computational systems consisting of many simple processing elements connected together to perform a particular task. Since the ANNs are powerful computational devices with capability of parallel processing, learning, generalization, and fault/noise tolerating, they provide powerful tools to solve the optimization problems. To use neural networks in optimization tasks, it is commonly needed to have a mathematical model of neural networks.

The PSO is one of the newly developed optimization techniques with many attractive feature and promising potentials. It was originally developed to create a computer model for simulating the social behavior of bird flocking, fish schooling, and swarming theory [2, 3]. The PSO algorithm shows that, with some modifications, the mentioned social behavior model can provide a powerful optimizer. The simplicity is one of the key attractive features of the PSO technique. It is a non-gradient and derivative-free method, which involves only two simple equations. However, like most metaheuristic methods, it is a problem-dependent solution method, and for every problem, the PSO parameters must be carefully tuned to ensure a high quality performance [4].

Genetic and evolutionary algorithms are nature-inspired optimization methods that can be advantageously used for many optimization problems. They imitate basic principles of life and apply genetic operators like mutation, crossover, or selection to a sequence of alleles. The sequence of alleles is the equivalent of a chromosome in nature and is constructed by a representation which assigns a string of symbols to every possible solution of optimization problems [5].

7.2 Artificial Neural Networks

Recent achievements on ANNs promote a great interest principally due to their capability to learn or to approximate well any arbitrary nonlinear functions, and their ability to use in parallel processing and multivariable systems [1]. These capabilities of such networks could be properly used to solve engineering optimization problems.

For many years, it was a dream of scientists and engineers to develop intelligent machines with a large number of simple elements such as neurons in biological organisms. McCulloch and Pitts [6] published the first systematic study of the ANNs. Then, a group of researchers combined these biological and psychological insights to produce the first ANN [7, 8], and further investigations in ANN continued by several pioneer researchers. The primary factors for the recent resurgence of interest in the area of neural networks are dealing with learning in a complex multi-layer network mathematical foundation for understanding the dynamics of an important class of networks [9].

The interest in ANNs comes from the networks' ability to mimic human brain as well as its ability to learn and respond. As a result, neural networks have been used in a large number of applications and have been proven to be effective in performing complex functions in a variety of fields including optimal control systems. Adaptation/learning is a major focus of ANN research that provides a degree of robustness to the ANN model.

An ANN consists of a number of nonlinear computational processing elements (neuron), arranged in several layers including an input layer, an output layer, and one or more hidden layers in between. Every layer usually contains several neurons, and the output of each neuron is usually fed into all or most of the inputs of the neurons in the next layer. The input layer receives input signals, which are then transformed and propagated simultaneously through the network, layer by layer [1].

7.2.1 Basic Element

The ANNs are modeled based on biological structures for information processing, including specifically the nervous system and its basic unit, the neuron. Signals are propagated in the form of potential differences between inside and outside of cells. A simplified scheme for a typical neuron is shown in Fig. 7.1a. Each neuron is composed of a body (*soma*), one *axon*, and multitude of *dendrites*. Dendrites bring signals from other neurons into the cell body. The cell body of a neuron sums the incoming signals from dendrites as well as the signals from numerous synapses on its surface. Once the combined signals exceed a certain cell threshold, a signal is transmitted through the axon. However, if the inputs do not reach the required threshold, the inputs will quickly decay and will not generate any action. The axon of a single neuron forms synaptic connections with many other neurons through some *terminal branches*. Cell nonlinearities make the composite action potential a nonlinear function of the combination of arriving signals.

According to a neuron biological structure, a simple artificial neuron (mathematical model) is depicted in Fig. 7.1b, which shows the basic element of an ANN with input vector $\mathbf{x} = [x_1, x_2, \ldots, x_n]^T$, weights $w_1, w_2, \ldots, w_n$, summation operator $\sum$, and an activation function $f(.)$. A summarized model with a single activation function is shown in Fig. 7.2 [1]. It consists of a threshold (or bias). The values w_i's are weight factors associated with each node to determine the strength of input vector $\mathbf{x}$. Each input vector is multiplied by the associated weight of the neuron connection. Depending upon the activation function, if the weight is positive, the resulted signal commonly excites the node output; whereas, for negative weights, it tends to inhibit the node output. The node's internal threshold, θ, weighted by w_0, is the magnitude offset that affects the activation of the node output y as follows:

$$y = f\left(\sum_{i=1}^{n} w_i x_i + w_0\theta\right). \tag{7.1}$$

The network shown in Fig. 7.2 is a simple computing element and was called the *perceptron*. The nonlinear cell function (activation function) can be selected according to the application. Sigmoid

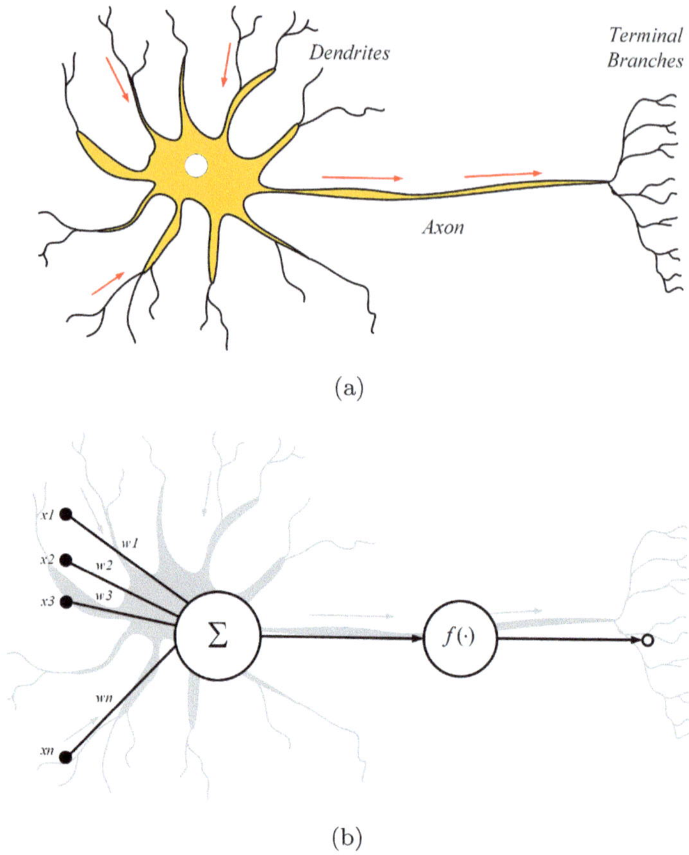

Fig. 7.1 Neuron. (**a**) Biological structure. (**b**) Corresponding artificial neuron

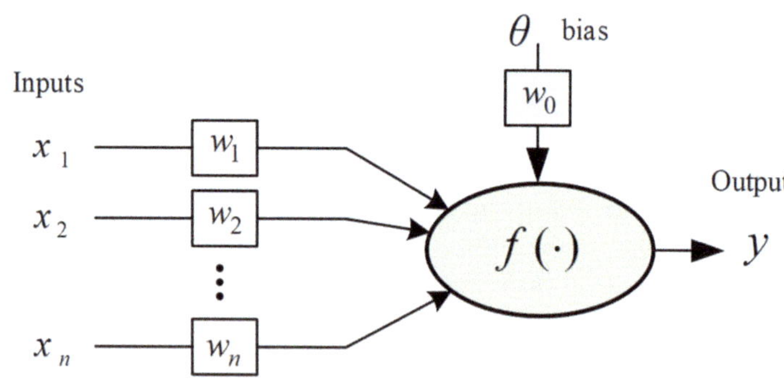

Fig. 7.2 Model of an artificial neuron

functions are a general class of monotonically non-decreasing functions taking on bounded values. It is noted that, as the threshold or bias changes, the activation functions may also shift.

Example 1 Assume a neuron with inputs **x**, weights **w**, and a unipolar sigmoid function $f(s) = \frac{1}{(1+e^{-s})}$. Calculate the neuron output for $\mathbf{x} = [0,\ 1,\ 0.5,\ 2]^T$, $\mathbf{w} = [1,\ 1,\ 0.5,\ 1.5]$, and $\theta = 0$.

Solution Summation of weighted inputs is $s = \mathbf{wx} = 4.25$, then the output can be achieved as $o = f(s) = f(4.25) = 0.9859$. □

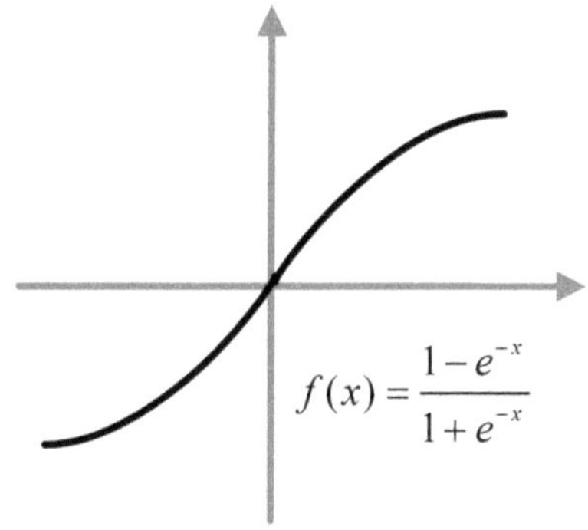

Fig. 7.3 Bipolar sigmoid function

7.2.2 Network of Artificial Neurons

As mentioned, an ANN consists of a number of neurons, arranged in several layers including an input layer, an output layer, and one or more hidden layers. A neuron accepts one or more input signals and produces one output, which is a nonlinear function of the sum of weighted inputs. The mapping from the input variables to the output variables can be fixed by setting all the weights associated with each neuron to some constants. In fact, the training of an ANN is a procedure to adjust these values so that the ANN can map all the input values to the corresponding output values [10].

In a multi-layer ANN, the neuron functions in the hidden and output layers are usually selected from nonlinear functions such as bipolar sigmoid function (Fig. 7.3), while linear neurons are used in the first layer. The number of hidden layers and the units in each layer is entirely dependent on the problem at hand, and there is no mathematical approach to obtain the optimum number of hidden layers [11], since such selection is generally fall into the application oriented category. The number of hidden layers can be chosen based on the training of the network using various configurations. Moreover, for many ANN training algorithms, such as backpropagation, the derivative of $f(.)$ is needed thus the selected activation function must be differentiable.

Example 2 Consider a three-layer ANN with inputs $\mathbf{x}$ and weights $\mathbf{W}_1$ and $\mathbf{W}_2$, as in Fig. 7.4. The neurons of the first layer use linear functions, while neurons in the hidden and the output layers use unipolar sigmoid function $f(s) = \frac{1}{(1+e^{-s})}$. Calculate the output vector of the ANN for

$$\mathbf{x} = \begin{bmatrix} 0.5 \\ -2 \\ 1 \end{bmatrix}, \mathbf{W}_1 = \begin{bmatrix} 1 & 0 & 0.5 \\ 0.2 & 1 & -1 \\ 2 & 0 & -3 \\ 1 & 2 & 0 \end{bmatrix},$$

$$\mathbf{W}_2 = \begin{bmatrix} 0.5 & 1 & 0 & 1 \\ 1 & 2 & 3 & 0.5 \end{bmatrix}.$$

Solution The summation of weighted inputs for the hidden layer is

$$\mathbf{s}_1 = \mathbf{W}_1 \mathbf{x} = \begin{bmatrix} 1 \\ -2.9 \\ -2 \\ -3.5 \end{bmatrix}$$

Fig. 7.4 A 3-layer ANN

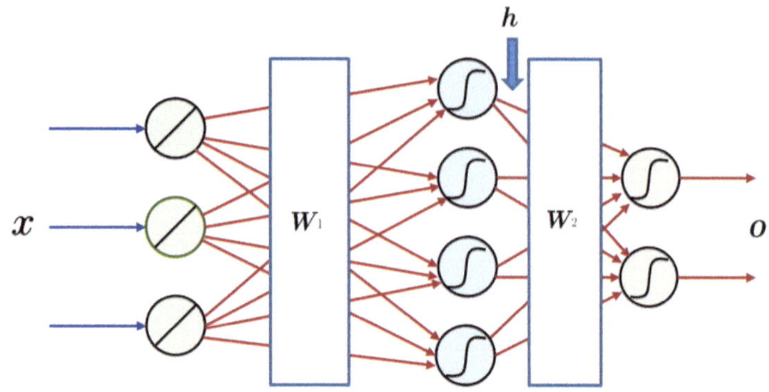

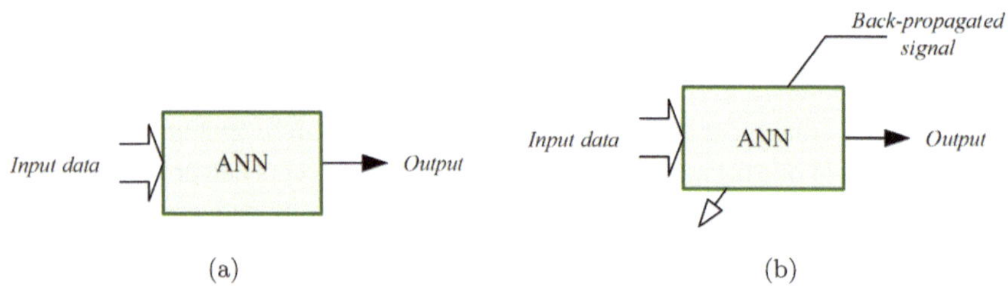

Fig. 7.5 Common ANN configurations. (**a**) Feedforward. (**b**) Backpropagation

and the output of the hidden layer is

$$\mathbf{h} = f(\mathbf{s}_1) = \begin{bmatrix} 0.7311 \\ 0.0522 \\ 0.1192 \\ 0.0293 \end{bmatrix}.$$

Then, summation of weighted inputs for the output layer is obtained as $\mathbf{s}_2 = \mathbf{W}_2\mathbf{h} = \begin{bmatrix} 0.4470 \\ 1.2076 \end{bmatrix}$. Finally, the ANN outputs through a feedforward calculation can be achieved as $\mathbf{o} = \begin{bmatrix} o_1 \\ o_2 \end{bmatrix} = f(\mathbf{s}_2) = \begin{bmatrix} 0.6099 \\ 0.7699 \end{bmatrix}$. □

The research on neural networks promotes a great interest principally due to the capability of static ANNs to approximate arbitrarily well any continuous function. The most used ANN structures are *feedforward* and *backpropagation (recurrent)* networks.

The above two configurations are presented in Fig. 7.5a, b, respectively. Backpropagation, which is a gradient descent learning algorithm, is one of the most popular learning algorithms in all mentioned configurations. It back propagates the error signals from the output layer to all the hidden layers, such that their weights can be adjusted accordingly. Backpropagation is a generalization of the least mean square (LMS) procedure for feedforward multi-layered networks with hidden layers. It uses a gradient descent technique, which updates the weights between neurons in its original and the simplest form by an amount proportional to the partial derivative of the error function with respect to the given weight [10].

7.2.3 Learning Process

Learning and *adaptation* are two keywords associated with the notion of ANNs. There exist different learning algorithms for ANNs, however, normally encounter some technical problems such as local minimization, low learning speed, and high sensitivity to initial conditions, among others. Recent advances in static and dynamic ANN have created a profound impact on the neural architecture for adaptation and optimal control, introduction to the backpropagation algorithms, identification, prediction, and optimization problems in dynamic systems/environments.

The backpropagation learning algorithm is known as one of the most efficient learning procedures for multi-layer ANNs. One reason for widely usage of this algorithm which will be described later is simplicity. This learning algorithm provides a special attribute for the design and operation of dynamic ANN for a given task such as control error minimization in complex dynamic systems. There are several approaches to derive the backpropagation algorithm. Direct analytic computation and recursive update techniques are two basic learning approaches to determine ANN weights. Learning algorithms may be carried out in continuous-time or in discrete-time via differential or difference equations for the weights. There are many learning algorithms, which can be classified into three categories: (1) *supervised learning* algorithm uses a supervisor that knows the desired outcomes and tunes the weights accordingly; (2) *unsupervised learning* algorithm uses local data, instead of supervisor, according to emergent collective properties; and (3) *reinforcement learning* algorithm uses some reinforcement signal, instead of output error of that neuron, to tune the weights [1].

Unlike the unsupervised learning, both supervised and reinforcement learning algorithms require a supervisor to provide training signals in different ways. In supervised learning, an explicit signal is provided by the supervisor to guide the learning process, while in reinforcement learning, the role of the supervisor is more evaluative than instructional [12]. On the other hand, the learning process in an ANN could be offline, or online. Offline learning is useful in feedforward applications such as classification and pattern recognition; while in optimal control applications, usually the online learning, which is more complex, is needed.

Selection of initial conditions in an ANN is also known as an important issue in optimization applications. Methods for the selection of initial conditions can be classified into three categories according to the form of the used information [13]: random initial conditions without use of the learning sample, deterministic initial conditions without use of the learning sample, and initial conditions with use of the learning sample.

The learning process of an ANN is usually realized through an optimization problem. For instance, it could be done by minimizing an error signal given by

$$E = \frac{1}{2} \|\mathbf{o} - \mathbf{o}_d\|_2^2 \tag{7.2}$$

where $\mathbf{o}_d$ represents the desired (reference) signal and $\mathbf{o}$ represents the output signal, both in $\mathbb{R}^K$ and indexed by k. It is desirable to find a set of parameters in the connection weights that minimizes the error E. It is useful to consider how the error varies as a function of the given connection weights in the system.

Implemented algorithm for updating weights is based on the backpropagation learning which is typically described for the ANN in Fig. 7.6. Let $\mathbf{x} \in \mathbb{R}^N$ be the input layer vector index by n, $\mathbf{h} \in \mathbb{R}^M$ be the hidden layer output vector indexed by m. Moreover, $\mathbf{W}_1 \in \mathbb{R}^{M \times N}$ and $\mathbf{W}_2 \in \mathbb{R}^{K \times M}$ be the weight matrices, where w_1^{mn} is the weight between neuron n in the input layer and neuron m in the hidden layer. Similarly, w_2^{km} is the weight between neuron m in the hidden layer and neuron k in the output layer. Vectors $\mathbf{net}_1 \in \mathbb{R}^M$ and $\mathbf{net}_2 \in \mathbb{R}^K$ are also indexed by m and k, respectively.

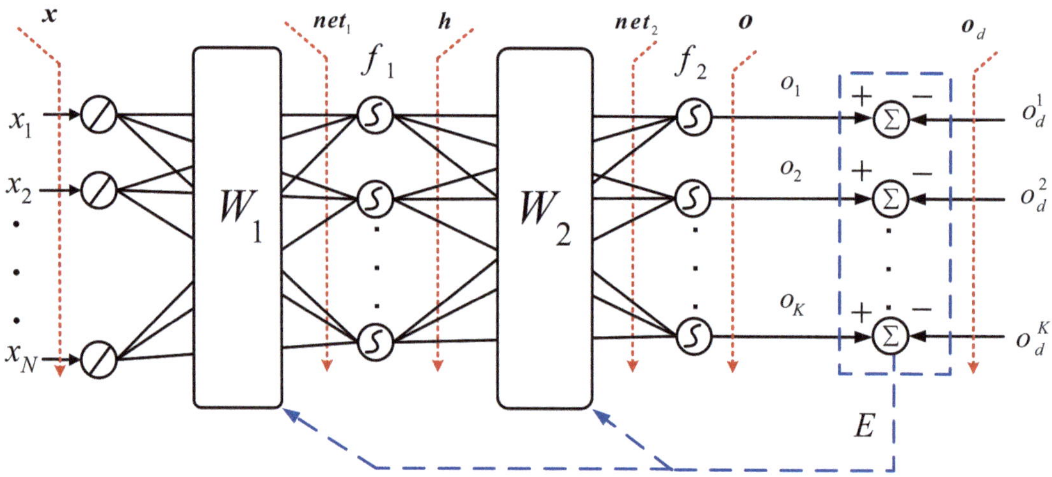

Fig. 7.6 A multi-layer ANN with backpropagation learning

To minimize the error signal in (7.2), beginning from initial weight matrices, these matrices are updated in an iterative manner as

$$\mathbf{W}_i(t+1) = \mathbf{W}_i(t) + \Delta\mathbf{W}_i, \quad i = 1, 2. \tag{7.3}$$

where $\Delta\mathbf{W}_i$ is the increment of $\mathbf{W}_i$. The mentioned procedure finds the values of all the connection weights that minimizes the error function using the gradient descent method. That is, after each pattern has been presented, the error gradient moves towards its minimum for that pattern, provided a suitable learning rate.

Learning of weights by employing the gradient descent method is done by computing $\Delta w_2^{km} = \eta\frac{\partial E}{\partial w_2^{km}}$ in $\mathbf{W}_2$ for all k and m, and $\Delta w_1^{mn} = \eta\frac{\partial E}{\partial w_1^{mn}}$ in $\mathbf{W}_1$ for all m and n, where $\eta \geq 0$ is a *learning rate* given by a small positive constant. Taking advantage of the chain rule,

$$\frac{\partial E}{\partial w_2^{km}} = \nabla_{\mathbf{o}} E^T D_{\mathbf{o}}(\mathbf{net}_2) D_{\mathbf{net}_2}(w_2^{km}) \tag{7.4}$$

where $\nabla_{\mathbf{o}} E$ is the gradient of E with respect to $\mathbf{o}$ that is equal to $(\mathbf{o} - \mathbf{o}_d)$. The second term, $D_{\mathbf{o}}(\mathbf{net}_2) = \mathbf{diag}[\frac{\partial o_1}{\partial net_2^1}, \ldots, \frac{\partial o_K}{\partial net_2^K}]$, is a diagonal matrix representing the Jacobian matrix of the transformation $\mathbf{o}(\mathbf{net}_2)$. Each entry $\frac{\partial o_k}{net_2^k} = f_2'(net_2^k)$, where $f_2(.)$ is the activation function in the output layer. The third term, $D_{\mathbf{net}_2}(w_2^{km}) \in \mathbb{R}^K$, is the partial derivative of $\mathbf{net}_2$ with respect to w_2^{km} computed as $D_{\mathbf{net}_2}(w_2^{km}) = [0, \ldots, h_m, \ldots, 0]^T$, where h_m is located at the kth entry. Substituting these derivations into (7.4), it is concluded that

$$\Delta w_2^{km} = \eta(o_k - o_d^k) f_2'(net_2^k) h_m. \tag{7.5}$$

Similarly,

$$\frac{\partial E}{\partial w_1^{mn}} = \nabla_{\mathbf{o}} E^T D_{\mathbf{o}}(\mathbf{net}_2) D_{\mathbf{net}_2}(\mathbf{h}) \times D_{\mathbf{h}}(\mathbf{net}_1) D_{\mathbf{net}_1}(w_1^{mn}) \tag{7.6}$$

where $\nabla_{\mathbf{o}} E$ and $D_{\mathbf{o}}(\mathbf{net}_2)$ are the same in (7.4). The third term, $D_{\mathbf{net}_2}(\mathbf{h}) = \mathbf{W}_2$, is the Jacobian of the transformation $\mathbf{net}_2(\mathbf{h})$. The forth term, $D_{\mathbf{h}}(\mathbf{net}_1) = \mathbf{diag}[\frac{\partial h_1}{\partial net_1^1}, \ldots, \frac{\partial h_M}{\partial net_1^M}]$, is a diagonal matrix where each entry $\frac{\partial h_m}{\partial net_1^m} = f_1'(net_1^m)$ and $f_1(.)$ is the activation function in the hidden layer. Finally, $D_{\mathbf{net}_1}(w_1^{mn}) \in \mathbb{R}^M$ is the partial derivative of $\mathbf{net}_1$ with respect to w_1^{mn} computed as $D_{\mathbf{net}_1}(w_1^{mn}) = [0, \ldots, x_n, \ldots, 0]^T$, where x_n is located at the mth entry. Substituting these derivations into (7.6), we derive

$$\Delta w_1^{mn} = \eta \sum_{k=1}^{K} (o_k - o_d^k) f_2'(net_2^k) w_2^{km} f_1'(net_1^m) x_n. \tag{7.7}$$

Example 3 Consider a water tank with a constant input and output flows (w) as shown in Fig. 7.7. Assume the following equation represents the thermodynamic behavior of the water tank system:

$$T(k+1) = fT(k) + (1-f)T_c(k) + \frac{1}{m}(1-f)q(k+1)$$

where the input and output temperatures are assumed to be T_c and T, respectively. q is the flame temperature which can be used as a control input, m shows the amount of available water in the tank, and $f = e^{-0.5\frac{m}{w}}$. Design a 3-layer ANN-based controller to fix the water temperature at 30 °C.

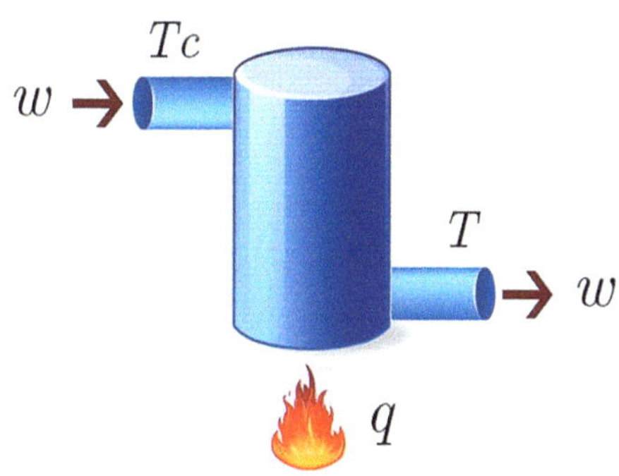

Fig. 7.7 Water tank example

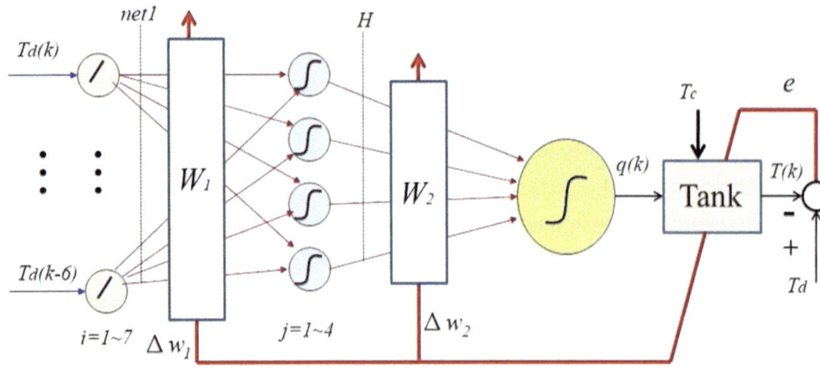

Fig. 7.8 ANN-based control framework for the water tank example

Solution Here, the optimization problem is considered to minimize the squared error between instant water temperature and the desirable one, i.e., to minimize $E = \frac{1}{2}(T_d - T)^2$.

To achieve the mentioned objective, the ANN control structure shown in Fig. 7.8 is proposed. As it can be seen, this network has seven units in the input layer, four units in the hidden layer, and one unit in the output layer. The number of output layer's neurons is equal to the number of control parameters that must be adjusted. The neural network acts as a controller to regulate the water temperature by minimizing the error signal. Here, the input vector of neural network is chosen as

$$\mathbf{x} = [T_d(k), T_d(k-1), T_d(k-2), T_d(k-3),$$
$$T_d(k-4), T_d(k-5), T_d(k-6)]^T.$$

As mentioned, linear functions are considered for the first layer, and for the second and third layers, the sigmoid functions are chosen as shown in Fig. 7.8. The main advantage of using these nonlinear functions is in performing a smooth update of the weights.

Selection of initial conditions in an ANN-based optimization system is also known as an important issue. In the multi-objective control problems, some initial values may not guarantee the achievement of objectives with a satisfactory value of optimization function. The initial conditions are usually selected according to the a priori information about distributions at the already-known structure of the feedforward ANN and selected optimization strategy. Here, the initial quantities in the applied ANN scheme (Fig. 7.8) are considered as $\eta = 0.15$, $\mathbf{W}_1 = 0.2 \times rand(4, 6)$ and $\mathbf{W}_2 = 0.1 \times [rand(1, 4) - 0.6]$. □

The following MATLAB codes could be used to a simple design of ANN-based optimal controller for the given example. Figure 7.9 shows that the proposed closed-loop system can effectively regulate the output temperature against any change in temperature of the input water.

```
%=============================================================
%——————————————————————— Initial weight setting:
clear;
clc;
w1old=rand(4,7)/5;
w2old=(rand(1,4)-.6)/10;
eta=0.15;
z=1:8000;
l(1)=12;
l(2)=18;
td=30*ones(size(z)); %Desired temp.
m=8.5;
w=2600;
deltat=2;
f=exp(-(m/w)*deltat);
d=(1/m)*(1-f);
%————————————————————————————Initial q & t setting:
for k=1:7;
q(k)=0;t(k)=l(1);
end;
deltaw1=0;
deltaw2=0;
%——————————————————————————Main loop:
for k=7:7000;
if k<=3500;
j=1;
else
j=2;
end;
tc=l(j)*ones(size(z));
%————————————————————————————————Update of weights:
w1new=w1old+deltaw1;
w2new=w2old+deltaw2;
w1old=w1new; w2old=w2new;
x=[td(k) td(k-1) td(k-2) td(k-3) td(k-4) td(k-5) td(k-6) ]';
% NN,s input vector
%—————————————————————————————————— Feed forward calculation:
net1=w1new*x;
h=1./(1.+(exp(1)).^(-net1));
net2=w2new*h;
q(k+1)=net2;
t(k+1)=f*t(k)+(1-f)*tc(k)+d*q(k+1);        % Plant model
m=1;
while abs((t(k+1)-t(k-m+1))/(q(k+1)-q(k-m+1)))>=1;
m=m+1;
end;
%———————————————————————————————— Error back propagation algorithm:
delta=(td(k+1)-t(k+1))*((t(k+1)-t(k-m+1))/(q(k+1)-q(k-m+1)))*1;
deltaw2=eta*delta*h';
sigmaj=delta*(h.*(1.-h)).*((w2new)');
deltaw1=eta*sigmaj*x';
end ;                              % end of main loop
%———————————————————————————————— Plot
w=1:k+1;
plot(delta*w,t);
grid on;
%=============================================================
```

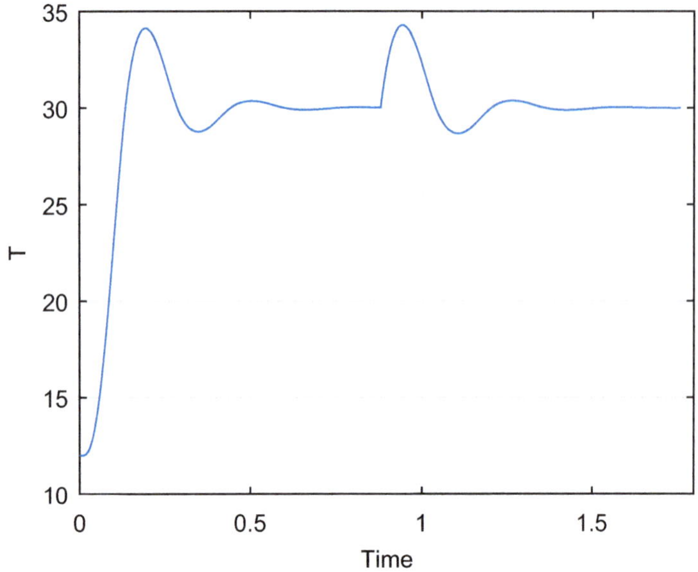

Fig. 7.9 System response to the change in the input temperature

7.2.4 An Application Example for Forecasting Photovoltaics Power and Load in a Microgrid

Classically, in order to optimize the power operation and cost in a photovoltaic (PV) system, power generation forecasting is needed. An ANN-based forecasting approach is given in [10]. In this approach, a stochastic process is applied to consider solar power generators and load uncertainty. Then, uncertainties of PV power generation and load demand are considered for calculating the MG operating power reserve.

A three-layer backpropagation ANNs is applied for the next 24 h PV power generation and load forecasting. The predicted results are used to predict the errors of forecasting and uncertainty analysis. Compared with conventional forecasting schemes, the ANNs have some additional advantages such as adaptability to new conditions and a high input data error tolerance. Many correlated factors such as solar irradiation, humidity, pressure, cloud cover percentage, temperature of ambient/panels, and wind speed may affect the output of PV power generators, different parameters have been used in the past works.

The overall system in the given example is an MG with 1 MW PV generator and 3 micro-gas turbines (MGTs) with 300 kW rated power (for each one) are considered. First, two backpropagation ANNs are used for PV power and load forecasting. Then, another two ANNs have been applied to forecast errors of PV power and load prediction. For the sake of error analysis, an hourly probability density function of all the predicted errors is used. Here, a backpropagation ANN for the next 24 h PV power generation prediction is developed based on last n-hours PV power generation, solar irradiation, and air temperature in the forecast day. The ANN structure is shown in Fig. 7.10. In the performed study, the PV power forecasts are evaluated with data from a real 17 kW PV power plant.

Firstly, the ANN is trained with historical record data and then predicted weather data are used as input variables of the designed ANNs to get hourly PV power generation output prediction. The efficiency of the proposed method is validated by analyzing the maximum error and mean absolute

Fig. 7.10 An ANN for PV power forecasting

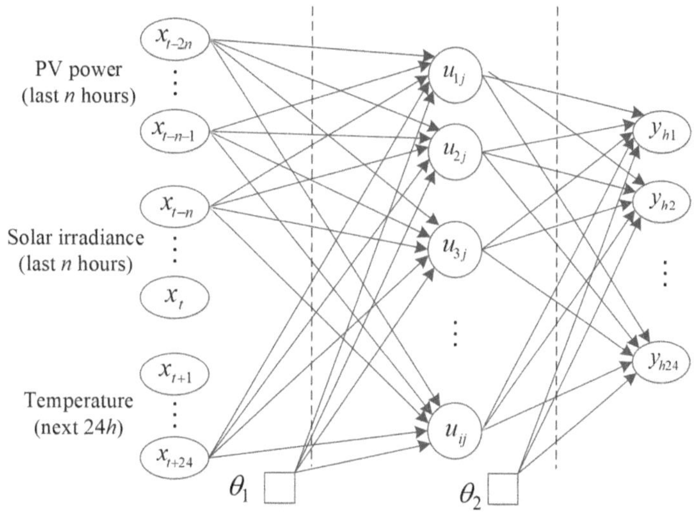

percentage error between predicted and measured values as well as standard deviations. The results are evaluated with data from a real 17 kW PV power plant. The input data for the first layer are last 24 hourly points of PV power and solar irradiation, as well as 24 points of hourly averaged temperature of the predicted day. A trial-and-error method has been used to determine the appropriate hidden layer number and the hidden neuron number in each hidden layer. A total of 24 hourly predicted points of PV power output are taken as output (output layer). 60%, 20%, and 20% of two-year real data are used to create the training set, the validation set, and the test set, respectively [10].

Similarly, an ANN is used for the load forecasting. The ANN is a three-layer ANN, which the input layer includes last n hours load demand measurements and 24 h average temperature of the forecast day; an output layer gives next 24 h load demand prediction and a hidden layer. Like PV power forecast, the real load data are used, and for ANN training, validation, and test, the amount of 60%, 20%, and 20% of the existing data is applied, respectively. The root mean square error (RMSE) has been minimized to calculate hidden neurons number and weights. The minimum RMSE is found with 370 hidden layer units and with last 48 h load measurements.

The forecasted values and actual measurements in a random day for both PV power and load are given in Fig. 7.11. Using (7.8), results of the RMSE and the mean absolute error (MAE) between the forecasted values ($\mathbf{y}_{pred}$) and actual values ($\mathbf{y}_{meas}$) for next 24 h are given in Table 7.1. The input historical load consumption of last 48 h is used, while other input variables belonged to the last 24 h.

$$\begin{aligned} RMSE &= \sqrt{\frac{1}{n}\sum_{i=1}^{n}\left(\mathbf{y}_{meas}(i) - \mathbf{y}_{pred}(i)\right)^2}, \\ MAE &= \frac{1}{n}\sum_{i=1}^{n}\left|\mathbf{y}_{meas}(i) - \mathbf{y}_{pred}(i)\right|. \end{aligned} \tag{7.8}$$

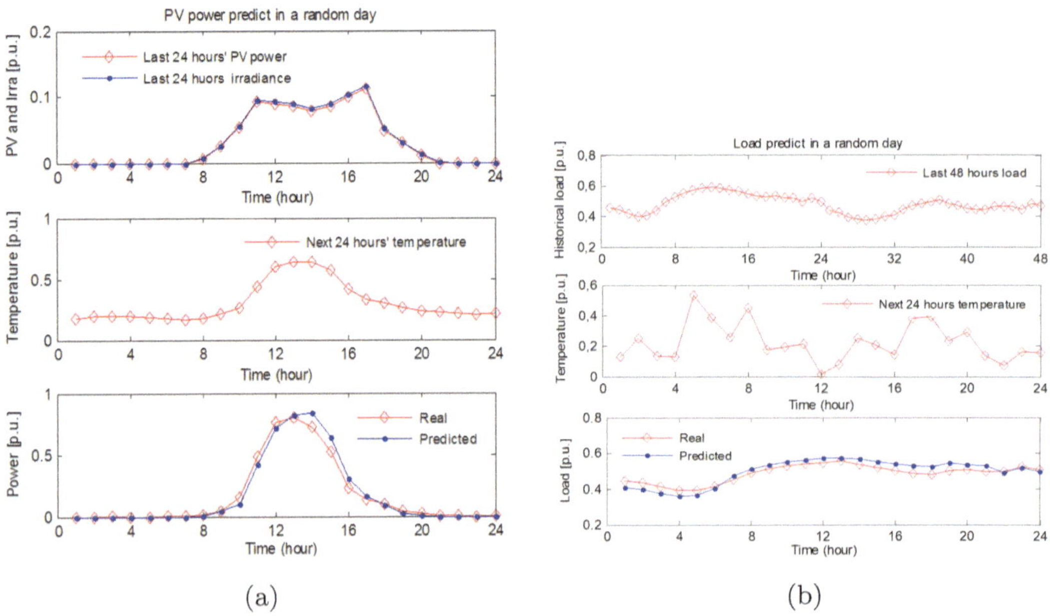

Fig. 7.11 ANN-based forecasting results in a random day. (**a**) PV power. (**b**) Load

Table 7.1 Errors of the PV power and load forecasting

	PV Power		Load	
Data	RMSE [%]	MAE [%]	RMSE [%]	MAE [%]
Training set	5.09	2.69	4.14	3.22
Validation set	5.58	3.13	4.75	3.77
Test set	5.95	3.12	4.90	3.84

7.3 Particle Swarm Optimization

Particle swarm optimization (PSO) is a population-based stochastic optimization technique. It belongs to the class of direct search methods that can be used to find a solution to an optimization problem in a search space. In the PSO method, a swarm consists of a set of individuals, known as *particles*. Each particle i is characterized by a *position* vector $\mathbf{x}_i(k) \in \mathbb{R}^n$ and a *velocity* vector $\mathbf{v}_i(k) \in \mathbb{R}^n$ at each time or iteration k. The position vector of every particle represents a potential solution to the under study optimization problem. In each iteration, the i-th particle flies from the current position to a new position using the velocity vector. A swarm consists of a number of particles or solution candidates flies through a feasible solution region to explore points where optimal solutions exist.

During their search, particles interact with each other in a certain way as to optimize their search experience. In each iteration, particles share their best position information with the rest of the swarm. Then, each particle updates its position based on its own and the swarm's best search experience. The best position for the i-th particle up to iteration k, represented by $\mathbf{x}_i^*(k)$, is determined according to the best value for the objective function. Indeed, $\mathbf{x}_i^*(k)$ is the best position that particle i achieved based on its own experience up to iteration k. Furthermore, the best position found by all particles in the population (global best position) up to iteration k is represented as $\mathbf{x}_g^*(k)$. It is the best particle position based on the overall swarm's experience.

In each iteration, the best particle position, global position, and the corresponding objective function values should be saved. For the next iteration, the position $\mathbf{x}_i(k)$ and velocity $\mathbf{v}_i(k)$ corresponding to the i-th particle can be updated using

$$\mathbf{v}_i(k+1) = w\mathbf{v}_i(k) + \left(c_1 r_1 \mathbf{x}_i^*(k) - \mathbf{x}_i(k)\right) + \left(c_2 r_2 \mathbf{x}_g^*(k) - \mathbf{x}_i(k)\right) \tag{7.9a}$$

$$\mathbf{x}_i(k+1) = \mathbf{x}_i(k) + \mathbf{v}_i(k+1) \tag{7.9b}$$

where w is the inertia weight and it decreases the number of iterations. r_1 and r_2 are random numbers in the interval [0 1] to introduce stochastic nature to the particles' movement. c_1 and c_2 are positive acceleration factors, and they keep balance between the particle's individual and social behavior.

The velocity vector in (7.9) consists of three terms to determine the next position: first term denotes the previous velocity, this is the stored velocity from the previous iteration to regulate each particle from making severe changes in its direction between consecutive iterations. Second term shows the cognitive component that represents the attraction force that each particle has towards its best position achieved based on its own flying experience. The last one is the social component, which corresponds to each particle tendency to be attracted towards the best position discovered among the entire individuals in a swarm [4].

To maintain a balance between the individuality and sociality, c_1 and c_2 are usually set to be equal. If c_1 is set greater than c_2, each particle individual performance will be weighted more in (7.9) and it is more likely that the algorithm will get trapped in local solutions (i.e., the best solution achieved by that individual particle). On the contrary, if c_1 is set less than c_2, that algorithm might fail to converge. The inertia weight parameter w introduced in (7.9) allows the velocity vector to start with larger values, and then it decreases as the iteration index increases to limit any big particle movements towards the end of the optimization process. This modification improves the convergence characteristics significantly [4]. A standard PSO algorithm includes the below steps [1]:

Step 1 all particles are initialized via a random solution. In this step, each particle position and associated velocity are set by randomly generated vectors. Dimension of position should be generated within a specified interval, and the dimension of velocity vector should also be generated from a bounded domain using uniform distributions.

Step 2 compute the objective function for the particles.

Step 3 compare the value of the objective function for the present position of each particle with the value of objective function corresponding to pre-specified best position, and replace pre-specified best position by the present position, if it provides a better result.

Step 4 compare the value of the objective function for the present best position with the value of the objective function corresponding to global best position, and replace present best position by the global best position, if it provides a better result.

Step 5 update the position and velocity of each particle according to (7.9).

Step 6 stop the algorithm if the stop criterion is satisfied. Otherwise, go to step 2.

7.3.1 An Application Example for Optimal Tuning of Droop Parameters in a Microgrid

Consider a microgrid (MG) with three inverter-based DGs as shown in Fig. 7.12a. Each DG is considered as a voltage source inverter (VSI), which its general block diagram is depicted in Fig. 7.12b

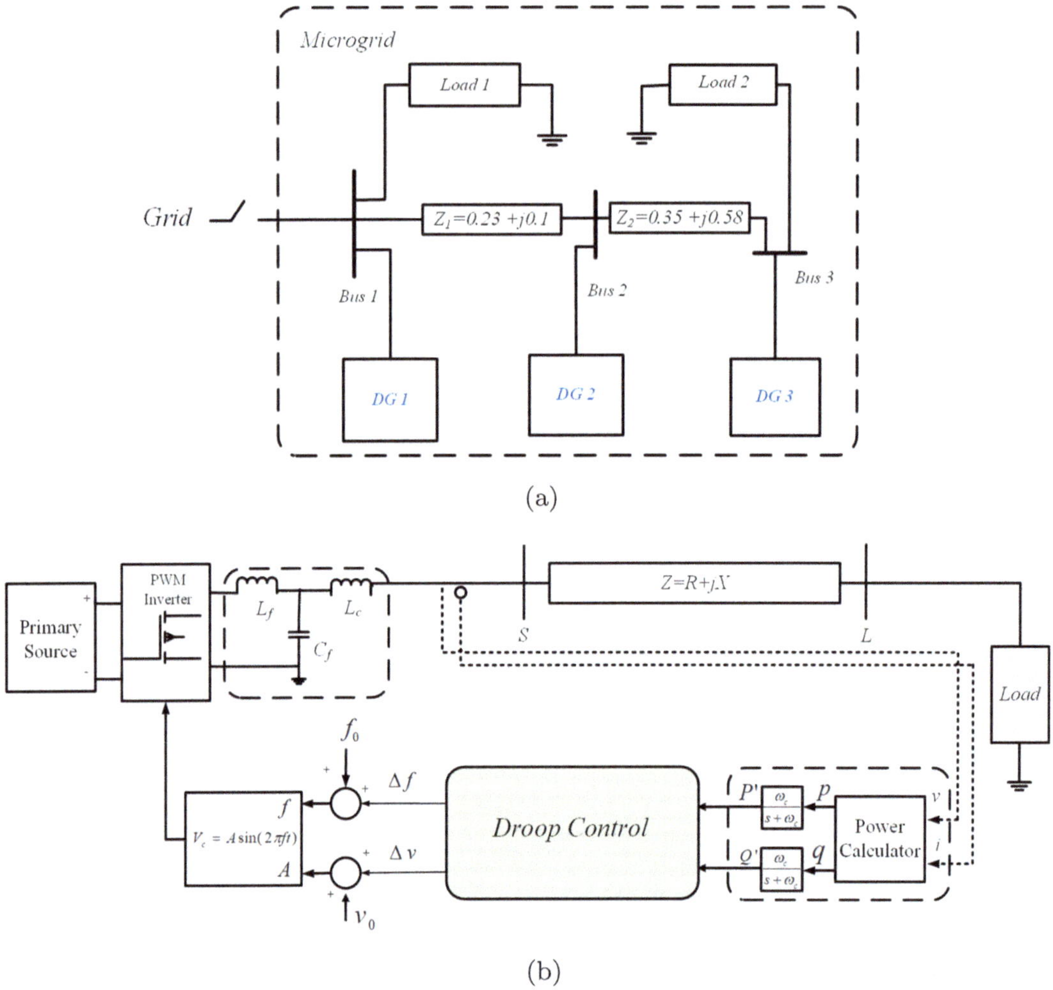

Fig. 7.12 An MG example with three DGs. (**a**) MG structure. (**b**) DG block diagram

[14]. The LCL output filter has been added to prevent the resonance impact in the output network. Also, it damps the distortion of output sinusoidal waveform and reduces high frequency harmonics caused by switching operations. In Fig. 7.12b, ω_c represents the cut-off frequency of low-pass filters. The power controller determines the real and reactive power, and renders P' and Q' to the *droop control* block to perform simultaneous voltage and frequency control according to the following dynamics [15]:

$$\Delta f = \frac{1}{K_f}\left(\frac{Z}{X}P' - P_0\right) + K_R K_V \Delta V_s + K_R Q_0 \tag{7.10}$$

$$\Delta V_s = \frac{1}{K_V}\left(\frac{Z}{X}Q' - Q_0\right) - K_R K_f \Delta f - K_R P_0 \tag{7.11}$$

where $K_R = \frac{R}{X}$, $K_f = -\frac{1}{k_p}$, $k_v = -\frac{1}{kq}$. The f, v_s, P_0, and Q_0 are inverter frequency, voltage, and rated active and reactive power, respectively.

The MG system parameters are given in [14]. Three DGs are connected to two local load banks, at bus 1 and bus 3. It is shown in [14] that in the MG, the droop controllers depend on the line parameters. Since, determining the instant amount of line parameters is difficult, to achieve a desirable response, one can consider a virtual R and X for each DG. Therefore, for the given case study there are six line parameters that must be determined ($[R_1\ X_1\ R_2\ X_2\ R_3\ X_3]$).

By properly estimating the virtual line parameters, the droop controllers exhibit high performance and acceptable response at different load change scenarios. Here, after defining an objective function to minimize the frequency and voltage fluctuations and using the PSO algorithm, the optimal virtual line parameters are estimated. Simultaneous control of frequency and voltage, as a main objective, can be presented in the following optimization problem:

$$\min_{\mathbf{x}}\ f(\mathbf{x}) = \sum_{i=1}^{n} \alpha_i |\Delta f_i(\mathbf{x})| + \beta_i |\Delta v_i(\mathbf{x})| \tag{7.12a}$$

$$\text{subject to } |\Delta f_i| \le E_f,\ |\Delta v_i| \le E_v,\ i=1,\ldots,n \tag{7.12b}$$

$$0.01^{\Omega} \le R_i \le 5^{\Omega},\ 0.01^{\Omega} \le X_i \le 0.3^{\Omega},$$

$$i = 1,\ldots,n. \tag{7.12c}$$

where $\mathbf{x}$ is the virtual line resistance (R) and reactance (X) of DGs ($\mathbf{x} = [R_1\ X_1\ R_2\ X_2\ R_3\ X_3]$). Δf_i and Δv_i are the frequency and voltage deviations of the i-th DG output, respectively. α_i and β_i are the stress coefficients of frequency and voltage in node i, respectively. n is the total number of DGs in the MG; E_f and E_v are the maximum allowable frequency and voltage deviations, respectively.

PSO Algorithm

Each particle is defined by $\mathbf{x}_i(k) = [R_1\ X_1\ R_2\ X_2\ R_3\ X_3]$. In initializing step, the particles are created with random positions and velocities. Following starting the algorithm, the position and velocity of all particles are made from previous step information. These particles are placed into the system and slip values of frequency and voltage are obtained from their nominal values. Here, the number of swarm particles is fixed at 5. The stop condition is met with maximum iteration of 20 epochs. The PSO algorithm is realized using the following MATLAB codes:

```
%==========================================================
%% Optimal Tuning of DGs Droop parameters using PSO
clear all;
warning off;
clc;
%% input data %
%% power controller
omega_c=60;
P0=0;
Q0=0;
%% Initial Simple Droop characteristics
```

```
mp=(0.15*2*pi)/1;
nq=(0.02)/2;
kq=nq;
kp=mp;
kf=-1/kp;
kv=-1/kq;
%% LCL parameters
lf1=6e-3;
rf1=0.2;
cf=50e-6;
lf2=3e-3;
rf2=0.1;
%% switichng frequency and sample time
fs=4000;
ts=1e-5;

%% Initialization
%initial line parameters
rl1=0.1;
xl1=0.1;
rl2=0.1;
xl2=0.1;
rl3=0.1;
xl3=0.1;

n = 5;              % Size of the swarm (no of birds)
bird_setp   =20;    % Maximum number of "birds steps"
dim = 6;            % Dimension of the problem

c2 =1.5;            % PSO parameter C1
c1 = 0.2;           % PSO parameter C2
w =0.8;             % pso momentum or inertia

fitness=0*ones(n,bird_setp);

%initialize the r parameter
R1 = rand(dim, n);
R2 = rand(dim, n);
current_fitness =0*ones(n,1);

%Initializing swarm velocities and position

current_position = (rand(dim, n));
velocity = .3*randn(dim, n) ;
local_best_position  = current_position ;

% Evaluate initial population

for i = 1:n
    current_fitness(i) = tracklsq(current_position(:,i));
    i
end

local_best_fitness  = current_fitness ;
[global_best_fitness,g] = min(local_best_fitness) ;

for i=1:n
    globl_best_position(:,i) = local_best_position(:,g) ;
    j=i
end
```

```
%  VELOCITY UPDATE
velocity = w *velocity + c1*(R1.*(local_best_position-current_position)) ...
+ c2*(R2.*(globl_best_position-current_position));

% SWARM UPDATE
current_position = current_position + velocity ;

%  evaluate a new swarm

%% Main Loop
iter = 0 ;        % Iterations' counter
while ( iter < bird_setp )
iter = iter + 1

for i = 1:n,
current_fitness(i) = tracklsq(current_position(:,i)) ;
end

for i = 1 : n
        if current_fitness(i) < local_best_fitness(i)
           local_best_fitness(i)  = current_fitness(i);
           local_best_position(:,i) = current_position(:,i);
        end
end

   [current_global_best_fitness,g] = min(local_best_fitness);

if current_global_best_fitness < global_best_fitness
   global_best_fitness = current_global_best_fitness;

    for i=1:n
        globl_best_position(:,i) = local_best_position(:,g);
    end

end

 velocity = w *velocity + c1*(R1.*(local_best_position-current_position)) ...
 + c2*(R2.*(globl_best_position-current_position));
 current_position = current_position + velocity;

 sprintf('The value of interation', iter );

end %

              xx=fitness(:,20);
              [Y,I] = min(xx);
              current_position(:,I)
%==========================================================
```

The function `tracklsq` can be written as follows:

```
%==============================================================
function F = tracklsq(line_parameters)

        r11 = line_parameters(1);
        x11 = line_parameters(2);
        r12 = line_parameters(3);
        x12 = line_parameters(4);
        r13 = line_parameters(5);
        x13 = line_parameters(6);

        % Compute function value
        sim('SimFile.mdl');

a1=100;
a2=150;
a3=0.5*a1;
a4=0.5*a2;
a5=a1;
a6=a2;
df1=df1dv1.signals(1,1).values(30000);
dv1=df1dv1.signals(1,2).values(30000);
df2=df2dv2.signals(1,1).values(30000);
dv2=df2dv2.signals(1,2).values(30000);
df3=df3dv3.signals(1,1).values(30000);
dv3=df3dv3.signals(1,2).values(30000);
F=a1*abs(df1)+a2*abs(dv1)+a3*abs(df2)+a4*abs(dv2)+a5*abs(df3)+a6*abs(dv3);
end
%==============================================================
```

As it is seen above, the MG structure should be implemented in a SIMULINK file entitled SimFile. For this purpose, the detailed MG parameters are given in [15].

Algorithm Results

The results after applying the PSO algorithm to the case study are shown in Fig. 7.13. This figure shows the motion trajectory of the following global best position:

$$\mathbf{x}^{gbest} = \mathbf{x}_g^* = [0.0118,\ 0.0107,\ 4.9736,\ 0.1372,\ 4.5040,\ 0.0327].$$

To test and verify the effectiveness of the estimated virtual line parameters, consider the outage of DG_1 from the MG system, in time duration of 0.4 to 0.6 s. Simulation results are shown in Fig. 7.14. After removing DG_1 at $t = 0.4$ s, other DGs are going to compensate the DG_1 absence and stop the voltage and frequency deviations. Figure 7.14 shows that the voltage and frequency indices remained stable under serious local load fluctuation, and the system frequency returns to nominal value, properly.

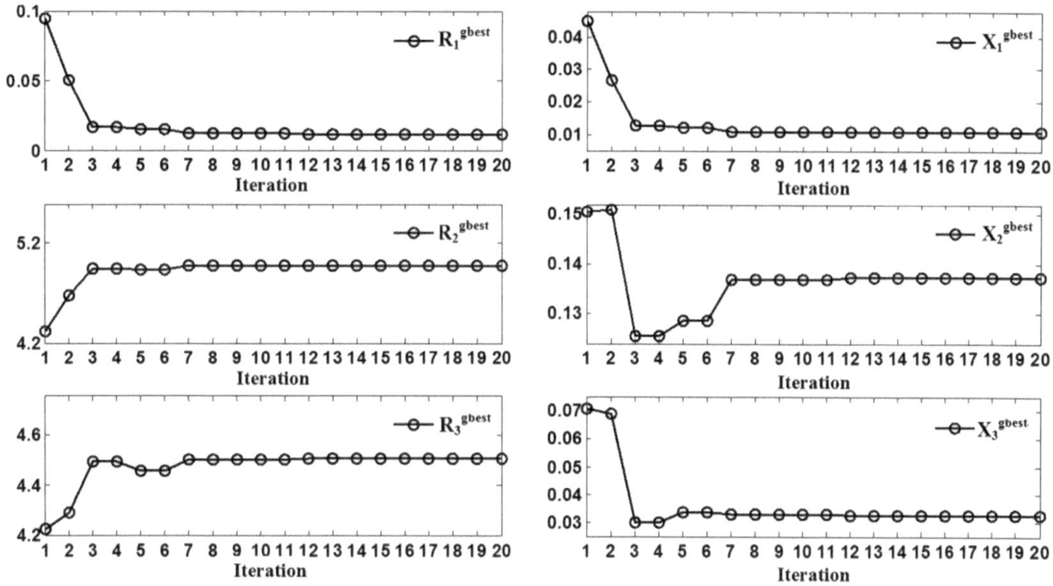

Fig. 7.13 The motion trajectory of global best positions for minimizing the defined objective function

7.4 Genetic Algorithm

7.4.1 An Overview

Genetic algorithms (GAs) rely on the analogy with the laws of natural selection and Darwin's most famous principle of survival of the fittest. The GA mechanism is inspired by the mechanism of natural selection where stronger individuals would likely be the winners in a competing environment. Normally in a GA, the parameters to be optimized are represented in a binary string, known as *chromose*. The *cost function* which determines the optimization problem objective or fitness function represents the main link between the system at hand and the GA unit, and also provides the fundamental source to provide the mechanism for evaluating of algorithm steps. To start the optimization, the GA uses randomly produced initial population or chromose set as solutions created by a random number generator. This method is preferred when a priori information about the problem is not available. There are basically three genetic operators used to produce a new generation. These operators are *selection*, *crossover*, and *mutation*. The GA employs these operators to converge at the global optimum. After randomly generating the initial population (as random solutions), the GA uses the genetic operators to achieve a new set of solutions at each iteration. In the selection operation, each solution of the current population is evaluated by its fitness, normally represented by the value of some objective function, and individuals with higher fitness values are selected [10].

Different selection methods such as stochastic selection or ranking-based selection can be used. In the selection procedure, the individual chromosomes are selected from the population for later recombination/crossover. The fitness values are normalized by dividing each one by the sum of all fitness values named *selection probability*. The chromosomes with higher selection probability have a higher chance to be selected for later breeding.

The crossover operator works on pairs of selected solutions with certain crossover rate. The crossover rate is defined as the probability of applying crossover to a pair of selected solutions, i.e.,

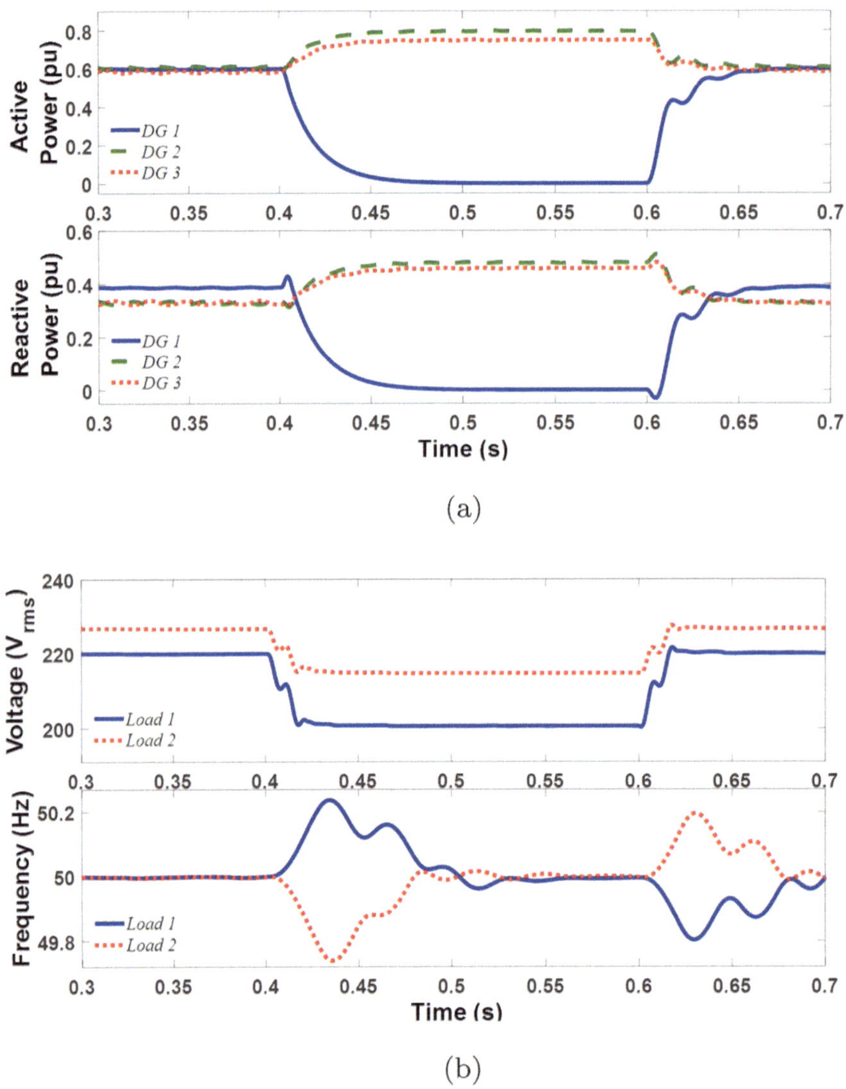

Fig. 7.14 System response due to outage of DG_1. (**a**) DGs active and reactive power. (**b**) Voltage and frequency on local loads

chromosomes. There are many ways to define the crossover operator. The most common way is called the *one-point crossover*. In this method, a point (e.g., for given two binary coded solutions of certain bit length) is determined randomly in two strings and corresponding bits are swapped to generate two new solutions. Furthermore, mutation is a random alteration with small probability of the binary value of a string position, and will prevent the GA from being trapped in a local minimum. Information generated by fitness evaluation unit about the quality of different solutions is used by the selection operation in the GA. The algorithm is repeated until a predefined number of generations have been produced [1]. A simple GA-based optimization approach is conceptually shown in Fig. 7.15a, and an overall flowchart for the GA is shown in Fig. 7.15b.

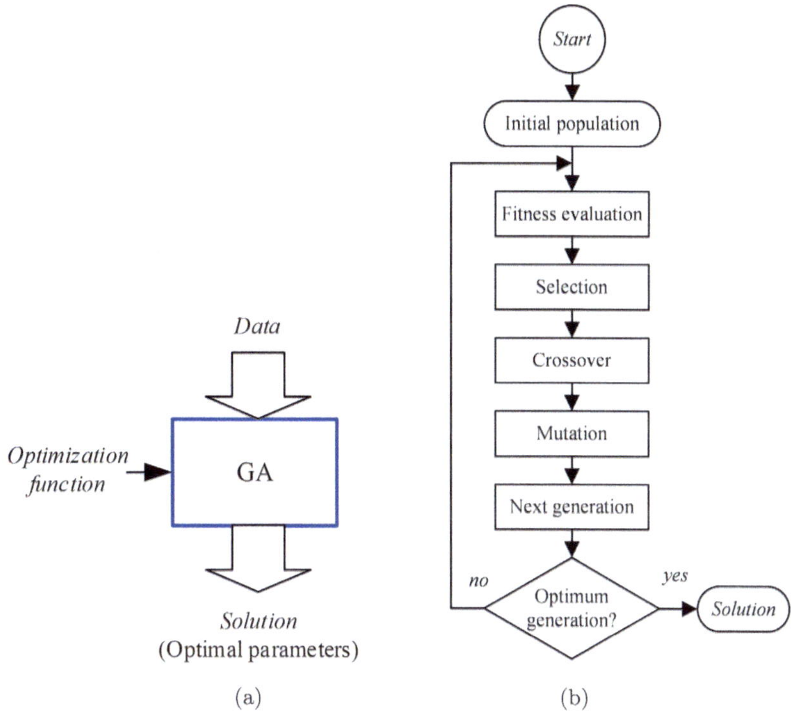

Fig. 7.15 GA-based optimization approach. (**a**) Overall scheme. (**b**) GA flowchart

Unlike the gradient-based optimization methods, the GAs operate simultaneously on an entire population of potential solutions (chromosomes or individuals) instead of producing successive iterates of a single element, and the computation of the gradient of the cost functional is not necessary. The GA is one of the rapidly emerging optimization approaches. The typical GA optimization steps can be summarized as follows:

Step 1 the initial population or chromosome set is generated (it may include numerous random binary strings). Then, the population (strings) may be decoded to the real numbers from domain of [0, 1].

Step 2 fitness values are calculated for each chromosome.

Step 3 the fitter ones are selected as parents.

Step 4 some pairs of parents are selected based on selection method and recombined to generate children.

Step 5 rarely, mutation is applied to the children. In other words, a few 0 bits flipped to 1, and vice versa.

Step 6 a new population is regenerated by allowing parents and children together.

Step 7 return to step 2 and repeat above steps until terminated conditions are satisfied.

Example 4 Consider a GA that uses pairing from top to bottom when selecting mates. Write a short GA codes in MATLAB, assume that the cost function must be provided by the user and converts the binary strings into useful variable values.

Solution Using the described GA steps it is easy to program a GA. A program sample could be written as follows [16]: □

```
%==========================================================
% initial population
N=200;                      % number of bits in a chromosome
M=8;                        % number of chromosomes must be even
last=50;                    % number of generations
sel=0.5;                    % selection rate
M2=2*ceil(sel*M/2);         % number of chromosomes kept
mutrate=0.01;               % mutation rate
nmuts=mutrate*N*(M-1);  % number of mutations

% creates M random chromosomes with N bits
pop=round(rand(M,N));   % initial population
for ib=1:last
cost=costfunction(pop);       % cost function

% ranks results and chromosomes
[cost,ind]=sort(cost);
pop=pop(ind(1:M2),:);
[ib cost(1)]
%mate
cross=ceil((N-1)*rand(M2,1));

% pairs chromosomes and performs crossover
for ic=1:2:M2
pop(ceil(M2*rand),1:cross)=pop(ic,1:cross);
pop(ceil(M2*rand),cross+1:N)=pop(ic+1,cross+1:N);
pop(ceil(M2*rand),1:cross)=pop(ic+1,1:cross);
pop(ceil(M2*rand),cross+1:N)=pop(ic,cross+1:N);
end

%mutate
for ic=1:nmuts
ix=ceil(M*rand);
iy=ceil(N*rand);
pop(ix,iy)=1-pop(ix,iy);
end %ic
end %ib
%==========================================================
```

7.4.2 GA for Multi-Objective Optimization

Initially, the majority of engineering optimization problems are inherently multi-objective problems, in that there are several conflicting objectives which need to be simultaneously achieved in the presence of determined constraints. If these objectives are analytically represented as a set of objective functions subject to the existing constraints, the optimization problem could be formulated as a multi-objective optimization problem. Mathematically, a multi-objective optimization (in form of minimization) problem can be expressed as

$$\min_{\mathbf{x}} \ (f_1(\mathbf{x}), f_2(\mathbf{x}), \ldots, f_M(\mathbf{x})) \tag{7.13a}$$

$$\text{subject to } \mathbf{x} \in \Omega \tag{7.13b}$$

where $\mathbf{x} \in \Omega$ is the vector of decision variables in the decision space Ω and $f_i(\mathbf{x})$ is the i-th objective function.

In a multi-objective problem, unlike a single optimization problem, the notation of optimality is not so straightforward and obvious. Practically in most cases, the objective functions are in conflict and show different behavior, so the reduction of one objective function leads to the increase in another. Therefore, in a multi-objective optimization problem, there may not exist one solution that is best with respect to all objectives. Usually, the goal is reduced to set compromising all objectives and determine a trade-off surface representing a set of non-dominated solution points, known as *Pareto-optimal* solutions, as stated in Definition 1.

Definition 1 In problem (7.13), the solution $\mathbf{x}_1$ dominates $\mathbf{x}_2$ ($\mathbf{x}_1$ is superior to $\mathbf{x}_2$) if

(i) $f_i(\mathbf{x}_1) \leq f_i(\mathbf{x}_2)$ for all $i = 1, 2, \ldots, M$.
(ii) $f_j(\mathbf{x}_1) < f_j(\mathbf{x}_2)$ for at least one j.

Moreover the solution $\mathbf{x}_i$ is said to be non-dominated or a Pareto-optimal point if there does not exist any $\mathbf{x}_j$ in the population that dominates $\mathbf{x}_i$.

A Pareto-optimal solution has the property that it is not possible to reduce any of the objective functions without increasing at least one of the other objective functions. Consequently, the solution of a multi-objective problem is not a single point, but a family of points as (the Pareto-optimal) solution set, which any member of the set can be considered as an acceptable solution. The choice of one solution over the others requires problem knowledge and a number of problem-related factors [1].

Practically, since there could be a number of Pareto-optimal solutions and the suitability of one solution may depend on system dynamics, environment, the designer's choice, etc., finding the center point of Pareto-optimal solutions set may be desired [1]. In the most common method, the solution is simply achieved by developing a population of Pareto-optimal or near Pareto-optimal solutions which are non-dominated. Non-dominated individuals are given the greatest fitness, and individuals that are dominated by many other individuals are given a small fitness. Using this mechanism, the population evolves towards a set of non-dominated, near Pareto-optimal individuals [17].

7.4.3 An Application Example for Load-Generation Balancing

The multi-objective GA methodology is conducted to balance load-generation via optimizing the load-frequency control loops in a multi-area power grid. The optimization objectives are summarized to minimize the area control error (ACE) signals in the interconnected control areas. To achieve this goal and satisfy an optimal performance, the parameters of the proportional-integral (PI) controller in the load-frequency control loop of each control area i can be selected through the minimization of

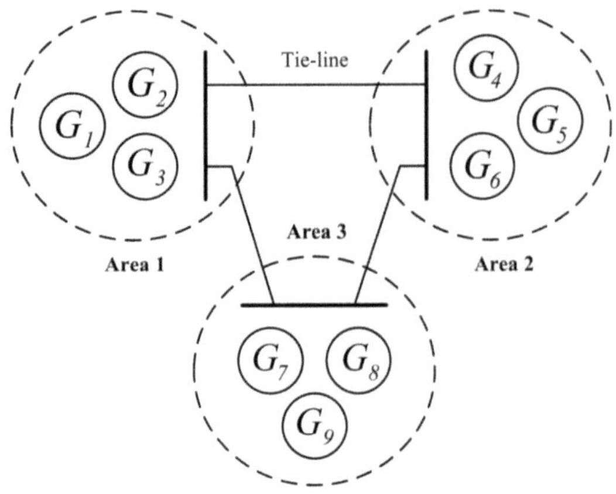

Fig. 7.16 Three-control area power system

$$f_i(\mathbf{x}_i) = \sum_{t=0}^{T} |ACE_{i,t}(\mathbf{x}_i)| \tag{7.14}$$

for all i, where T is equal to the simulation sampling time (sec), $|ACE_{i,t}|$ is the absolute value of ACE signal for area i at time t, and $\mathbf{x}_i$ is the vector of the corresponding PI controller parameters. A PI controller has two gain parameters. Therefore, using multi-objective GA optimization technique to tune the PI controllers, $M \times 2$ gains must be determined, where M is the number of control areas.

As mentioned above, the population of a multi-objective GA is composed of dominated and non-dominated individuals. The basic line of the algorithm is derived from a GA, where only one replacement occurs per generation. The selection phase should be done, first. Initial solutions are randomly generated using a uniform random number of PI controller parameters. The crossover and mutation operators are then applied. The crossover is applied on both selected individuals, generating two childes. The mutation is applied uniformly on the best individual. The best resulting individual is integrated into the population, replacing the worst ranked individual in the population.

Example 5 Consider a 3-area power system as shown in Fig. 7.16. Using appropriate definitions and state variables, the state space realization of control area i can be obtained as [14]

$$\dot{\mathbf{x}}_i = \mathbf{A}_i\mathbf{x}_i + \mathbf{B}_{1i}\mathbf{w}_i + \mathbf{B}_{2i}u_i \tag{7.15a}$$

$$y_i = \mathbf{C}_{yi}\mathbf{x}_i \tag{7.15b}$$

where $\mathbf{x}_i = [\Delta f_i,\ \Delta P_{tie-i},\ \mathbf{x}_{mi},\ \mathbf{x}_{gi}]^T$, $\mathbf{x}_{mi} = [\Delta P_{m1i},\ \Delta P_{m2i},\ \ldots,\ \Delta P_{mni}]$, $\mathbf{x}_{gi} = [\Delta P_{g1i},\ \Delta P_{g2i},\ \ldots,\ \Delta P_{gni}]$, $u_i = \Delta P_{C_i}$, $y_i = ACE_i = \beta_i \Delta f_i + \Delta P_{tie,i}$, and $\mathbf{w}_i = [\Delta P_{Li},\ v_i]^T$.

Here, Δf is frequency deviation, ΔP_m is governor valve position, ΔP_{gi} denotes the governor valve position change, ΔP_C is control action, ΔP_{tie} is net tie-line power flow, β is frequency bias, v is area interface disturbance, $\Delta P_{Li}(s)$ is load disturbance, and

(continued)

$\mathbf{A}_i = \begin{bmatrix} \mathbf{A}_{i11} & \mathbf{A}_{i12} & \mathbf{A}_{i13} \\ \mathbf{A}_{i21} & \mathbf{A}_{i22} & \mathbf{A}_{i23} \\ \mathbf{A}_{i31} & \mathbf{A}_{i32} & \mathbf{A}_{i33} \end{bmatrix}$, $\mathbf{B}_{1i} = \begin{bmatrix} \mathbf{B}_{1i1} \\ \mathbf{B}_{1i2} \\ \mathbf{B}_{1i3} \end{bmatrix}$, $\mathbf{B}_{2i} = \begin{bmatrix} \mathbf{B}_{2i1} \\ \mathbf{B}_{2i2} \\ \mathbf{B}_{2i3} \end{bmatrix}$, $\mathbf{A}_{i11} = \begin{bmatrix} -D_i/2H_i & -1/2H_i \\ 2\pi \sum_{j=1,j\neq i}^{N} T_{ij} & 0 \end{bmatrix}$, $\mathbf{A}_{i12} = \begin{bmatrix} 1/2H_i & \dots & 1/2H_i \\ 0 & \dots & 0 \end{bmatrix}_{2\times n}$, $\mathbf{A}_{i22} = -\mathbf{A}_{i23} = -\text{diag}[1/T_{t1i},\ 1/T_{t2i},\ \dots,\ 1/T_{tni}]$,

$\mathbf{A}_{i33} = \text{diag}[-1/T_{gli},\ -1/T_{g2i},\ \dots,\ -1/T_{gni}]$, $\mathbf{A}_{i31} = \begin{bmatrix} -1/(T_{g1i}R_{li}) & 0 \\ \vdots & \vdots \\ -1/(T_{gni}R_{ni}) & 0 \end{bmatrix}$,

$\mathbf{A}_{i13} = \mathbf{A}_{i21}^T = \mathbf{0}_{n\times 2}$, $\mathbf{A}_{i32} = \mathbf{0}_{n\times 2}$, $\mathbf{B}_{1i1} = \begin{bmatrix} -1/2H_i & 0 \\ 0 & -2\pi \end{bmatrix}$, $\mathbf{B}_{li2} = \mathbf{B}_{li3} = \mathbf{0}_{n\times 2}$,

$\mathbf{B}_{2i1} = \mathbf{0}_{n\times 2}$, $\mathbf{B}_{2i2} = \mathbf{0}_{n\times 1}$, $\mathbf{B}_{2i3} = [\alpha_{1i}/T_{g1i},\ \alpha_{2i}/T_{g2i},\ \dots,\ \alpha_{ni}/T_{gni}]^T$, and $\mathbf{C}_{yi} = [\beta_i,\ 1,\ \mathbf{0}_{1\times n},\ \mathbf{0}_{1\times n}]$.

Furthermore, H is equivalent inertia constant, D is equivalent damping coefficient, T_{ij} is tie-line synchronizing coefficient between areas i and j, R is droop characteristic, and α is participation factor. All parameters are given in Table 7.2. Assume in each control area, a PI controller provides an appropriate control input to minimize the ACE signal as

$$u_i = \Delta P_{ci} = k_{P_i} ACE_{i,t} + k_{Ii} \int ACE_{i,t} d\tau. \quad (7.16)$$

Find optimal PI parameters using GA.

Solution The system frequency response model for the 3-area power system is built in MATLAB/SIMULINK environment as shown in Fig. 7.17 [14]. The described multi-objective GA is applied to the 3-control area power system example and the closed-loop system response for simultaneous load step increase in three areas ($\Delta P_{L_1} = 100$ MW; $\Delta P_{L_2} = 80$ MW; $\Delta P_{L_3} = 50$ MW) is examined. The results for area 2 are shown in Fig. 7.18 and it is compared with the obtained results from application of robust H_∞-PI control methodology (using iterative LMI) which addressed in [18]. □

Table 7.2 Applied data for the given power system

Parameters	Genco								
MAV_{base} (1000 MW)	1	2	3	4	5	6	7	8	9
Rate (MW)	1000	800	1000	1100	900	1200	850	1000	1020
B_i (pu/Hz)	0.3483	0.3473	0.3180	0.3827	0.3890	0.4140	0.3692	0.3493	0.3550
D_i (pu MW/HZ)	0.015	0.014	0.015	0.016	0.014	0.014	0.015	0.016	0.015
R_i (Hz/pu)	3.00	3.00	3.30	2.7273	2.6667	2.50	2.8235	3.00	2.9412
$2H_i/f_0$ (pu.sec)	0.1677	0.120	0.200	0.2017	0.150	0.196	0.1247	0.1667	0.187
T_{ti} (sec)	0.4	0.36	0.42	0.44	0.32	0.40	0.30	0.40	0.41
T_{gi} (sec)	0.08	0.06	0.07	0.06	0.06	0.08	0.07	0.07	0.08
α_i	0.4	0.4	0.2	0.6	0	0.4	0	0.5	0.5
Ramp rate (MW/min)	8	8	4	12	0	8	0	10	10

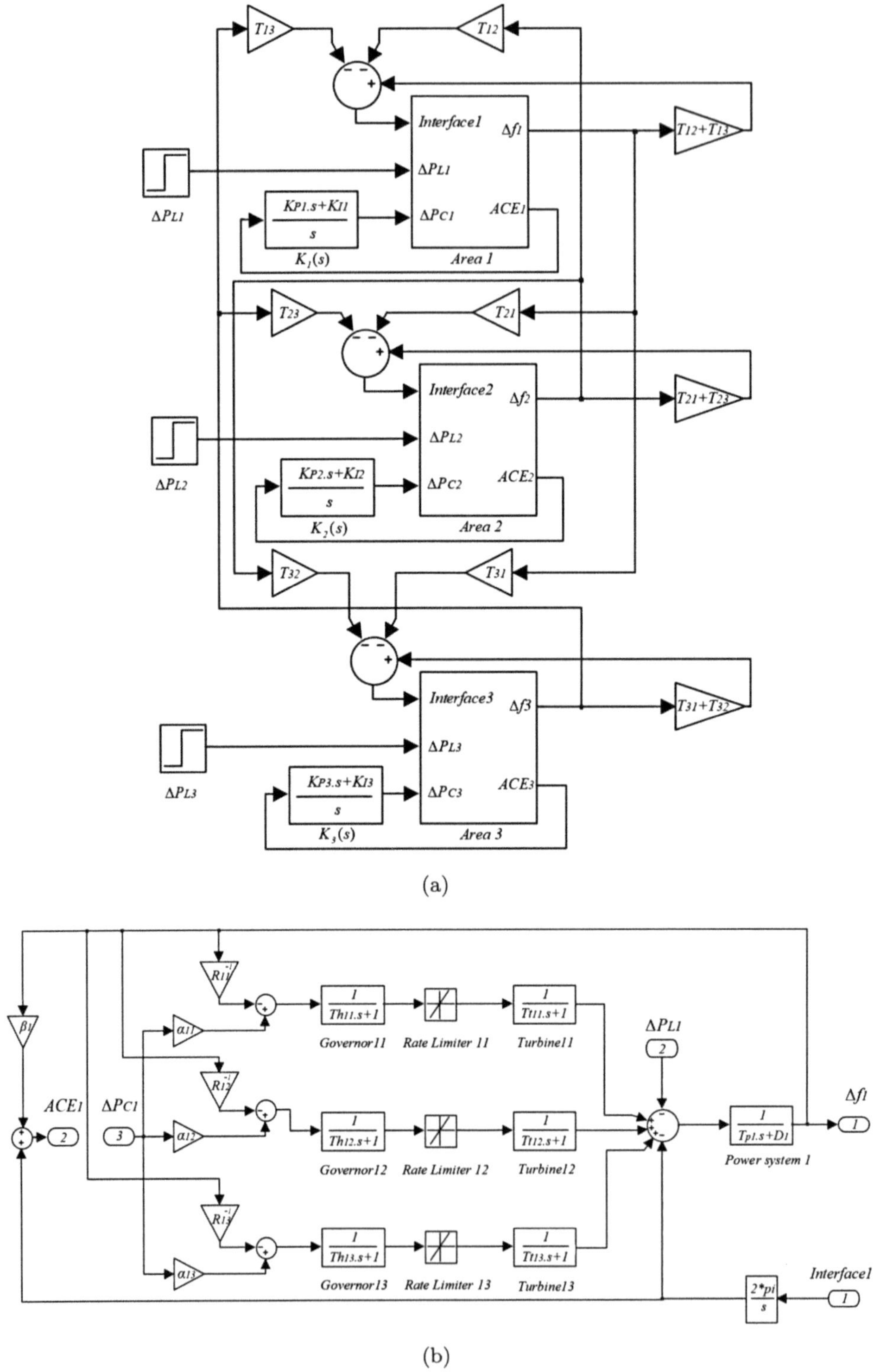

Fig. 7.17 Building the LFC model in SIMULINK environment. (**a**) 3-control area. (**b**) Detailed model of area 1

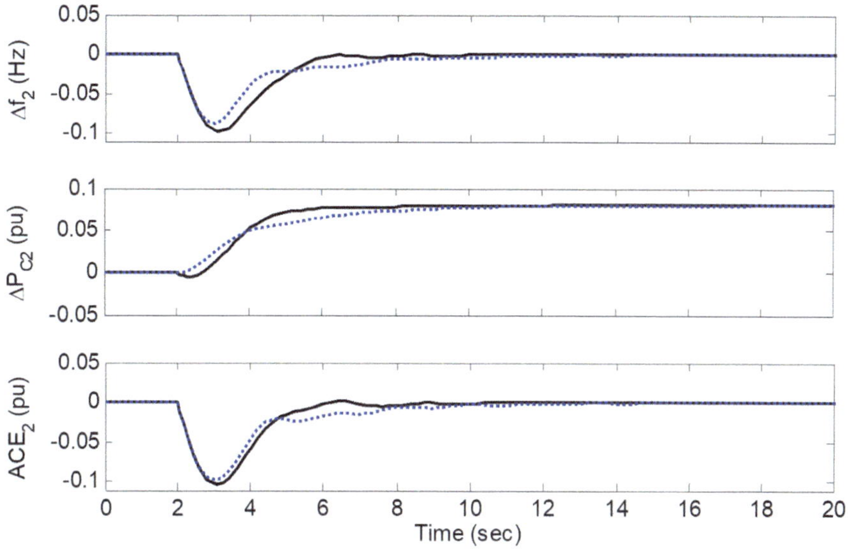

Fig. 7.18 System response in area 2 (solid line: GA methodology, dotted line: robust PI control)

7.5 Summary

Using the artificial intelligence and nature-inspired searching methods it is possible to overcome some limitations of traditional optimization approaches, and to increase the number of solvable optimization problems. The application of AI and EAs to many optimization problems often results good performance and high quality solutions.

While many of these approaches are still under investigation, however, numerous research works indicate the applicability of these approaches on the optimization issues. In this chapter, the application of ANNs, PSO, and GA as powerful techniques in optimization problems is explained. The relevant fundamentals are emphasized and several application examples are presented. It is shown that for successful and efficient use of AI techniques and EAs, it is not enough to simply apply the standard approaches. In addition, it is necessary to find a proper representation as well as understanding the given problem.

7.6 Problems

Problem 1 Solve Example 2 using bipolar sigmoid functions and discuss on the obtained results.

Problem 2 Consider the given four layers ANN in Fig. 7.19, with inputs $\mathbf{x}$ and weights $\mathbf{W}$. The neurons of the first layer use linear functions, while neurons in hidden and output layers use unipolar sigmoid function $f(s) = \frac{1}{(1+e^{-s})}$. Calculate the outputs of the ANN.

$$\mathbf{x} = \begin{bmatrix} 0.5 \\ -2 \\ 1 \end{bmatrix}, \mathbf{W}_1 = \begin{bmatrix} 1 & 0 & 0.5 \\ 0.2 & 1 & -1 \\ 2 & 0 & -3 \\ 1 & 2 & 0 \end{bmatrix}, \mathbf{W}_2 = \begin{bmatrix} -1 & 0.5 & 2 & 1 \\ 3 & -2 & -1 & -0.5 \\ -1 & 1 & 2 & 2 \end{bmatrix}, \mathbf{W}_3 = \begin{bmatrix} 0.5 & 1 & 0 \\ 1 & 2 & 3 \end{bmatrix}$$

Problem 3 Consider a two layers ANN with inputs $\mathbf{x} = [1,\ 0.5,\ 2,\ -1]^T$ and $d = -1$ in Fig. 7.20. After four iterations in the backpropagation learning algorithm, $\mathbf{W}(4) = [2,\ 4,\ 3,\ 1]$ is given. Find $\mathbf{W}(3)$ and $\mathbf{W}(5)$.

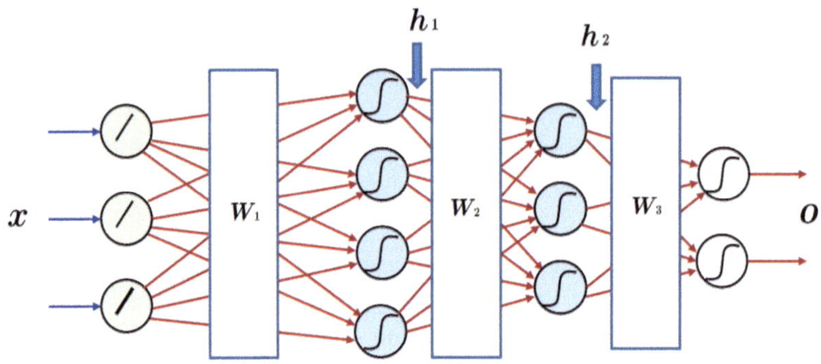

Fig. 7.19 Four layers ANN

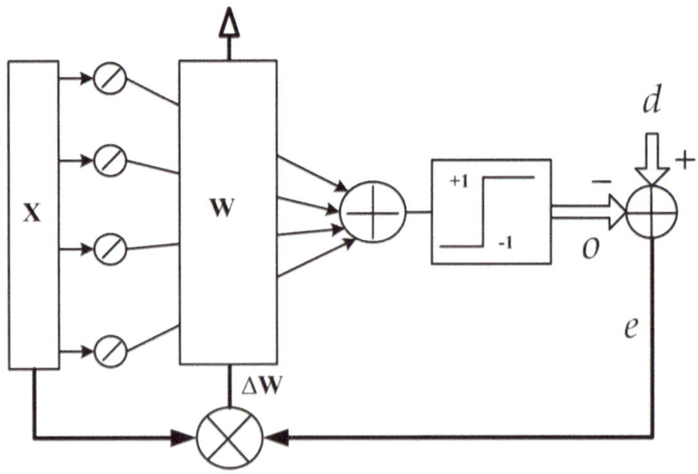

Fig. 7.20 Two layers ANN

Problem 4 Determine True and False statements:

(a) Hidden-layer units provide a better prediction of desired outputs for new inputs that it was not trained with.
(b) In backpropagation learning, we should start with a small learning parameter and slowly increase it during the learning process.
(c) A three-layer network with 5 neurons in each layer has a total of 50 connections and 50 weights.
(d) In GA, the chromosomes are coded in binary strings.
(e) Some conflicts among training exemplars in a NN can be resolved by adding features to the input vectors and adding input layer neurons to the network.
(f) Backpropagation tries to minimize the error of a NN by changing the connection weights.
(g) Using GA, the computation of the gradient of the cost functional is necessary.

Problem 5 Consider the ANN in Fig. 7.21 with the backpropagation learning algorithm using error gradient decent. If $\mathbf{x} = [1,\ 3,\ -1]^T$, $d = 1$ and $f(net) = \frac{1}{1+e^{-net}}$, find $\mathbf{V}$, $\mathbf{W}$, $\Delta\mathbf{V}$ and $\Delta\mathbf{W}$ after the first iteration.

Problem 6 In the following minimization problem with two criteria, find the necessary and sufficient conditions such that $\mathbf{x}_1$ dominates $\mathbf{x}_2$.

$$\min_{\mathbf{x}} \quad (f_1(\mathbf{x}), f_2(\mathbf{x}))$$

$$\text{subject to } \mathbf{x} \in \Omega$$

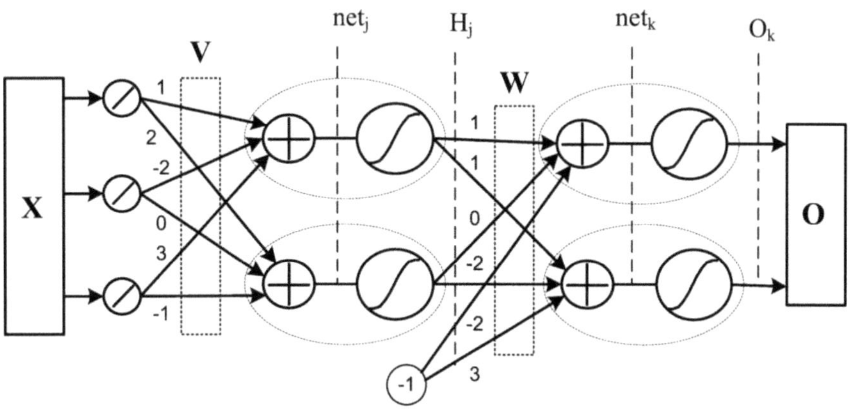

Fig. 7.21 Backpropagation learning algorithm

Problem 7 Using GA algorithm, solve the multi-objective minimization problem

$$\min_{x} \; (x^2, (x-2)^2)$$

$$\text{subject to } x \in [-6, 6]$$

Problem 8 Using GA algorithm, solve the following multi-objective minimization problem:

$$\min_{x} \; (f_1(\mathbf{x}), f_2(\mathbf{x}))$$

$$\text{subject to } x_1 \in [-5, 5], x_2 \in [-5, 5]$$

where $f_1(\mathbf{x}) = (x_1 - 2)^2 + (x_2 - 4)^2$ and $f_2(\mathbf{x}) = (x_1 - 3)^2 + (x_2 - 1)^2$.

Problem 9 Reconsider Example 4 with different population sizes and mutation rates. Which combination seems to work best? Explain.

Problem 10 Develop a PSO algorithm for a simple optimization example (e.g., the example given in Problem 7), and test it.

References

1. H. Bevrani, T. Hiyama, *Intelligent Automatic Generation Control* (CRC Press, Boca Raton, 2011)
2. J. Kennedy, R. Eberhart, Particle swarm optimization, in *Proceedings of the IEEE International Conference on Neural Networks*, vol. 4 (1995), pp. 1942–1948
3. R. Eberhart, J. Kennedy, New optimizer using particle swarm theory, in *Proceedings of the Sixth International Symposium Micro Machine and Human Science*, pp. 39–43, 1995
4. M.R. AlRashidi, M. AlHajri, A. Al-Othman, K. El-Naggar, Particle swarm optimization and its application in power systems, in *Computational Intelligence in Power Engineering* (Springer, Berlin, 2010), pp. 295–324
5. F. Rothlauf, *Representations for Genetic and Evolutionary Algorithms* (Springer, New York, 2006)
6. W.W. McCulloch, W. Pitts, A logical calculus of the ideas immanent in nervous activity. Bull. Math. Biophys. **5**(4), 115–133 (1943)
7. F. Rosenblatt, *Principles of Neurodynamics* (Spartan Press, Washington, DC, 1961)
8. J.L. McClelland, D.E. Rumelhart, *Parallel Distributed Processing Explorations in the Microstructure of Cognition*, vol. 2 (MIT Press, Cambridge, MA, 1986)
9. A. Zilouchian, M. Jamshidi, *Intelligent Control Systems Using Soft Computing Methodologies* (CRC Press, Boca Raton, 2001)
10. H. Bevrani, B. Francois, T. Ise, *Microgrid Dynamics and Control* (Wiley, New York, 2017)
11. H. Bevrani, T. Hiyama, Y. Mitani, K. Tsuji, M. Teshnehlab, Load-frequency regulation under a bilateral LFC scheme using flexible neural networks. Eng. Intell. Syst. J. **14**(2), 109–117 (2006)
12. F.L. Lewis, J. Campos, R. Selmic, *Neuro-Fuzzy Control of Industrial Systems with Actuator Nonlinearities* (SIAM Books, Philadelphia, 2002)

13. A.I. Galushkin, *Neural Networks Theory* (Springer, New York, 2007)
14. H. Bevrani, *Robust Power System Frequency Control* (Springer, New York, 2009)
15. H. Bevrani, S. Shokoohi, An intelligent droop control for simultaneous voltage and frequency regulation in islanded microgrids. IEEE Trans. Smart Grid **4**(3), 1505–1513 (2013)
16. R.L. Haupt, S.E. Haupt, *Practical Genetic Algorithms*, 2nd edn. (Wiley, New Jersey, 2004)
17. C.M. Fonseca, P.J. Fleming, Multiobjective optimization and multiple constraint handling with evolutionary algorithms—part I: a unified formulation. IEEE Trans. Syst. Man Cybern. **28**, 26–37 (1995)
18. H. Bevrani, Y. Mitani, K. Tsuji, Robust decentralised load-frequency control using an iterative linear matrix inequalities algorithm. IEE Proc. Gener. Transm. Distrib. **3**(151), 347–354 (2004)

Index

M. Fathi, H. Bevrani, *Optimization in Electrical Engineering*,
https://doi.org/10.1007/978-3-030-05309-3

MIX
Papier aus verantwortungsvollen Quellen
Paper from responsible sources
FSC® C105338

If you have any concerns about our products,
you can contact us on
ProductSafety@springernature.com

In case Publisher is established outside the EU,
the EU authorized representative is:
Springer Nature Customer Service Center GmbH
Europaplatz 3, 69115 Heidelberg, Germany

Printed by Libri Plureos GmbH
in Hamburg, Germany